AGRICULTURAL COOPERATIVES
Their Why and Their How

AGRICULTURAL COOPERATIVES
Their Why and Their How

GLYNN McBRIDE

**Department of Agricultural
Economics
Michigan State University
East Lansing, Michigan**

AVI PUBLISHING COMPANY, INC.
Westport, Connecticut

TO

My parents—one instilled a love for
learning—the other the ability to listen
and to hear.

My uncle—for making movement toward a
professional career financially possible.

My students in FSM 443—they inspired,
they challenged—they insisted—they
elicited enthusiasm.

My wife, Mary Jo, and son, David Glynn,
who made it all worthwhile.

Copyright 1986 by
THE AVI PUBLISHING COMPANY, INC.
P.O. Box 831
250 Post Road East
Westport, Connecticut 06881

Library of Congress Cataloging-in-Publication Data

McBride, Glynn.
    Agricultural cooperatives.

    Bibliography: p.
    Includes index.
    1. Agriculture, Cooperative—United States.
2. Agriculture, Cooperative—Law and legislation—United
States.   I. Title.
HD1491.U5M37   1986   334'.683'0973   86-17324
ISBN 0-87055-534-0

Printed in the United States of America
A B C D 5 4 3 2 1 0 9 8 7 6

Contents

Part II
THE "HOW" OF AGRICULTURAL COOPERATIVES   85

6 Capper–Volstead Corporations and Other Types of Business—Cooperative Principles   87

7 Economic Feasibility of a Cooperative   101

8 Cooperative Management Trio— Members, Directors, and Manager · 119

Preface

The genesis of this book dates back a number of years to an annual meeting of the American Institute of Cooperation. Cooperative leaders at that meeting openly expressed their feelings that they would like a much stronger commitment on the part of our land grant universities in particular and other educational institutions in general to creating a greater understanding of cooperatives and their role in our economy. Since I was in agreement with this position, the course in general group action which I was teaching was changed to emphasize the role of agricultural cooperatives. The need for a textbook designed to help in this emphasis became apparent. This book is the culmination of an effort to meet that need.

This book has been prepared with the student and instructor in mind. It is based upon experience gained from many years of interaction with students in a classroom setting, in discussions with them after class hours, and after they had finished their degree requirements and had taken positions in industry, government, or elsewhere. While objectives of the book have remained relatively unchanged over time, the substance and format as means of meeting those objectives have changed from time to time as a result of these contacts and discussions. The input of students over the years as reflected in this product is gratefully acknowledged.

The final format of the book, along with suggestions in regard to the substantive material it should contain, were strongly influenced by an event that occurred on the first day of class a few years ago. After the usual first-day elaboration of objectives, procedures, expectations, and hopes, a rather frail and meek-appearing student raised her hand and said, "Dr. McBride, I don't believe farm cooperatives are justified. I don't think they were in the beginning and I don't think they are now."

She was intelligent, articulate, and obviously sincere. She came from a nonrural, nonfarm background and was reflecting a position based on that experience. She had had no exposure to the agricultural area except as she shopped for food at the supermarket. She suggested strongly that the position she was taking represented the position of many people in our society.

The die was cast for the format of the book. The student was asked to wait until near the end of the course for a response to her statement. It was pointed out to her and the class that the first part of the course would be directed at an exploration of the economic rationale for the existence of the institutional arrangement known as the farm cooperative. The importance of facing up to the "why" of farm cooperatives was vividly demonstrated by the very sincere remarks of this student. The first part of the book would be devoted to this area—the why of farm cooperatives.

After this exploration in which the students are encouraged to develop a position regarding the why of farm cooperatives with which they feel comfortable, efforts move to the "how" aspect of farm cooperatives.

Efforts in both areas, the why and the how, are directed and guided by relevant questions. What is the structure of agriculture and of those industries with which it is most closely related—the providers of inputs to agriculture and those who buy the output of agriculture? What is the relevance of structure to marketing and for meeting specifications for products which consumers are demanding? What is the public interest involved in these areas and are the economic interests of the firms in various structural arrangements likely to be in accord with or divergent from the general public interest? If they are found to diverge, what then? Is agriculture unique on either the supply side or on the basis of the demand for the product it produces?

The how portion of the effort suggested in the book's format is also directed by certain relevant questions.

How does this form of business organization differ from other forms in our society? What is the relevance of these differences to the role of the members in their cooperative—to the role of management—to the role or area of financing? What are the issues and challenges faced by cooperative leaders as they use this form of business organization in trying to achieve their objectives?

The basic objectives of the student as encouraged and helped by the instructor would be aimed in the why section of the course toward reaching a comfortable and defensible position regarding whether farm cooperatives are economically justified. Once such a position is reached, the student is encouraged to develop experitse in articulating it in a professional manner and be prepared to do so on any appropriate occasion.

If the position taken is that farm cooperatives are justified, a most logical next step would appear to be to become knowledgeable in those areas relating to enhancing the probability that this institutional arrangement will, in fact, bring forth the performance of which it is potentially capable.

All efforts in the course, in the classroom, in discussions, in term papers, in the workbook assignments—everywhere—would be handled in a constructively critical setting. Movement toward professionalism on the part of the student is expected and is encouraged.

It is suggested that any student of industrial organization, whether in formal or informal educational settings, would have an interest in this book, especially the why of agricultural cooperatives. This would include those who, for any reason, would like help in establishing their own position regarding this business arrangement. It would also include those who are responsible for enacting legislation of any form as it relates to agriculture in general and to agricultural cooperatives in particular. Finally, it would include everyone who seeks help with questions relating to what is involved in the concept of the public interest and the conditions under which the performance of firms and industries are most likely to be in harmony with that interest.

This book is designed for use in a one-semester or one-quarter course in agricultural economics, agricultural marketing, agribusiness, and other related areas. It would typically be used by upper-level undergraduate students who had acquired some knowledge of economic principles and were exploring various forms of business organization as career possibilities.

The questions at the end of each chapter have multiple purposes. The focus questions are designed to encourage concern with details regarding the areas covered in the chapter. In addition, and perhaps more importantly, other questions are designed to encourage the student to see the larger picture of what is involved and to automatically start pulling together all relevant elements from many sources in developing that picture.

A Workbook of Cases in Cooperative Marketing is prepared as a supplement to the textbook. As indicated in its Preface, it is designed to provide students with practical and interesting problems in the cooperative area. In working through the problems in a hands-on manner, students become personally involved in the practical world of agricultural cooperative marketing.

The basic thrust throughout the book is toward the excitement and motivation of students and a facilitative role for enthusiastic teachers. The end product, hopefully, is a much greater understanding of the economic rationale and legal underpinning of this institutional arrangement and of the potential role it can play in helping to solve many of our pressing problems in marketing.

As is always the case, one incurs indebtedness in an effort such as this. Dr. Larry Connor, Head, Department of Agricultural Economics, Michigan State University, has been supportive of the effort in both tangible and intangible ways. Peggy Crawford was very helpful in the early days of the work in typing parts of the manuscript. Lori Ramirez graciously performed this very valuable role in later stages. The role performed by Jeanette Barbour in the later stages in editing, storage in the word processor, revising, and rearranging has been of inestimable value. Not once did she show signs of being annoyed or frustrated with the editing, reediting, and final editing which were necessary to get to the final form.

Further acknowledgment is made to my students who have contributed so much to this project. They continued to provide new, fresh, and helpful ideas along with motivation to prepare this book. Jack Evans and Michael Fassler read the manuscript and made suggestions.

Jack Barnes and Glenn Lake served as role models of cooperative practitioners. They were aware of and maintained that fine-line balance between a position of concern only with bottom-line consideration and one of recognizing the special uniqueness of this business form and using it judiciously in capturing its potential strength. Jack's efforts, along with those of Harold Lein, Charles Bucholtz, and our cooperatives, in making an endowed cooperative scholarship available for our students are a major contribution to future cooperative leadership. I, along with future recipients of the scholarship, thank them.

Anonymous reviewers and many consultants have contributed in special ways. Colleagues in my department, in other departments, and many professionals in other universities have provided encouragement to pursue this effort. George Dike has been helpful in many ways.

Dept capital obligations of this nature cannot be discharged or repaid in any of the usual ways. They can, however, be evidenced by permanent IOUs which extend into perpetuity. I gratefully provide that evidence of an annuity which is payable forever.

Part I
THE "WHY" OF AGRICULTURAL COOPERATIVES

Our efforts in this section will be aimed at developing, in a professional manner, a position in regard to whether agricultural cooperatives are justified. The economic rationale for the existence of this type of business organization will be explored in attempting to develop a position with which we feel comfortable, can articulate, and are willing to defend.

1

Marketing and Structure— How Related to Group Action

Before we move directly into considering the "why" of agricultural cooperatives, let us review general economic activity and why it takes place. What are the objectives of the economic actors and how do they attempt to meet their objectives? Are the objectives of the decision makers who guide economic activity of firms in complete alignment with so-called public interest objectives? If the results of economic activity indicate a divergence between the goals and objectives of the decision makers and those which the public would prefer, what are the implications for public policy in general and policy regarding agricultural cooperatives in particular? Finally, what is marketing and how is it involved in all our deliberations?

MARKETING AND ECONOMIC ACTIVITY

Marketing can be considered only in relation to or as a part of economic activity. Economic activity involves combinations and recombinations of resources on a planned basis. More than one decision maker is involved. The combinations and recombinations are made with ends or purposes in mind. The ultimate end is a product or service that meets the specifications of the potential user. To the extent that the ultimate objective is not met, resources may be wasted, efficiency may suffer, and transaction costs will be increased.

Most of the planning which results in the combination and recombination of resources is left to the market system in a free enterprise society. This does not mean that the Smithian Wand is operative, as we shall see later, but it means that most decisions are made by individuals and groups in pursuing their firm's goals. Governmental regulations and state planning are held to a minimum.

The ideal situation would be one in which each of the firms in the agricultural marketing system was concerned only with the overall task of getting products from the farm to the consumer and that all actions and activities were carried out with this overall perspective operating as a guiding force. This variation of the magic wand would come close to providing the products consumers wished to have and could very well be in accordance with the public interest.

This, however, is not the case. Each firm pursues its individual goals. Whether or not these goals and the results of the combinations and recombinations are consistent with efficiency, speed, and safety in the marketing of farm and food products which might be assumed as societal goals depends upon the internal economics of the firm itself and the structure of the industry of which the firm is a part. Both of these will be examined later in considering the role of marketing and the place of farm cooperatives in our economy.

VIEWS OF MARKETING

There is no generally agreed upon definition of marketing, although many working definitions are found. Each one perhaps reflects the perspective of the writer and of those whose views parallel those of the writer.

Those with an agricultural orientation usually stress the functional aspects of the activities involved in movement of a farm-produced raw material from the farm to the ultimate consumer. Functions such as production, hauling, processing, storage, wholesaling, and retailing are emphasized. Efficiency from the standpoint of location, transportation, capacity, seasonality, etc. is a major consideration. The perfect market concept serves as a model.

This marketing perspective encourages concern with individual functions and little, if any, attention is paid to their linkage in some sort of alliance to combine and recombine resources in a meaningful manner. It encourages the establishment and perpetuation of a production–marketing dichotomy. It has provided a foundation for the position taken by many farmers that the farming task is one of production and what happens outside the boundaries of the farm fence is the responsibility of others. No coordination in the decision-making process at any stage is necessary. It is simply a—this is production, that is marketing—posture. While this view is structurally based, the unique characteristics of the farm commodities involved should discourage its continued acceptance.

It has been pointed out by many writers that the marketing of agricultural products is different in several ways from marketing other products. One of the differences is tied in with the fact that many agricultural products keep their generic names all the way from the farm to the consumer. The hindquarter of a hog, for example, starts out as ham and the consumer buys ham. This same relationship holds for most farm products.

This is not the case with products of industries structured differently than agriculture. We rarely buy a truck or a tractor—we buy a GM or a Ford or a Deere or an International Farmall. At first, these differences may not appear to be significant, but upon further thought, they are basically grounded in the structural arrangements of the industries involved and are significant from the standpoint of conduct options available to the firms in the industries and from the consumer behavior area.

Another perspective of marketing could be called the "business school" view. In this view, marketing is developing a product and convincing potential consumers that it should be purchased. This view is grounded in the area of competitive structure and accepts the position that use of some resources is aimed at persuading consumers and influencing demand. Again, no resource allocation function or coordination in the decision-making process at any stage is provided.

As stated in the beginning, economic activity involves combining and recombining resources in a meaningful manner, which suggests a coordination and synchronizing of decisions at each stage of activity or function in such activities as will result in a product or service which is in conformity with the requirements of a consumer. This may be the ultimate consumer or it may be at some stage in the process. In any event, the activity involves doing whatever is necessary in making a raw material available and in shaping it to fit specifications of various users.

This suggests a coordinating and synchronizing force over this process of economics in action. An allocative function, especially at the early stages of the process, is also suggested. Neither of the above perspectives of marketing provides for such a role.

An approach, called the structural approach, has been accepted by many economists. As described by Shaffer, this approach to marketing includes the "system of markets and related institutions which organize the economic activity of the food and fiber sector of the economy."

The distinctive feature of this perspective is that it permits coordination of economic activity through institutional arrangements and is not entirely dependent upon the market for control over the economy. A major place is left for market prices to perform their traditional function, but there is a recognition that structural arrangements in most sectors are far removed from the perfect market concept and that government may have a legitimate role in taking steps which would be helpful in permitting the market to perform more satisfactorily. This perspective recognizes that in the absence of judiciously selected and applied institutional arrangements, the market may perform very poorly because of the competitive structure which has come to prevail in many sectors.

THE MARKETING UMBRELLA

The essence of the structural approach may be captured in what might be referred to as the marketing system umbrella (Fig. 1.1).

The umbrella reflects the functions that must be performed in combining and recombining resources from the raw material stage to other stages. It suggests, however, a decision-making process at each stage, with decisions relating to combinations that will result in a product at each stage. This product meets requirements at that stage and calls for further recombinations to be made. The process continues until the product or service, a result of economic decisions which have been made, is ready for the final consumer. If those decisions have been soundly based, the product or service will meet the specifications of the consumer. This ultimate matching signals that the decisions have, in fact, been soundly based.

Heavy reliance is placed upon economic intelligence at each of the decision-making points. Alternative combinations of resources are technically possible and alternate means to achieve the desired results must be worked out. Production techniques must be assessed in relation to engineering efficiency, input–output ratios, and so on. In addition,

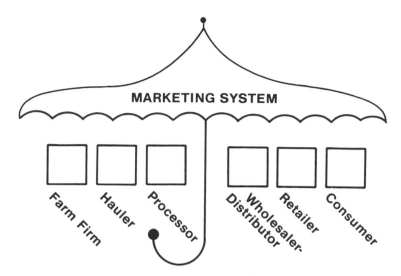

FIG. 1.1. The marketing system umbrella.

other variables such as environmental pollution concerns will have to be considered. Laws and regulations relating to safety, use of labor, and the like will be a part of the process.

This process takes place at each stage where combinations and recombinations of resources are considered. If the data provided for the decision-making process are adequate, the technical coefficients of production will suggest the most feasible production technique for combining resources in such a way that the output will have the planned capacity to create or satisfy consumer wants.

At each point where recombinations take place, a production process is activated that is based on technical coefficients of production and the demand factors which are determined to be most feasible. This means that production takes place at many points and part of the decision-making process has taken into account the economics of the additional functions such as transportation, storage, and warehousing of both the raw material, accessory materials, and the product in its planned form for the next consumer. The production–marketing dichotomy is thus shattered, since the whole process is lodged in a decision-making matrix and those functions which are necessary to the goals of the firm are performed at any place and time.

The umbrella concept encourages a diagnostic or clinical approach to problem-solving and can be an invaluable aid to the clinician. An assessment of the data and the outcome of the use of the data at each decision point in relation to product suitability or other relevant criteria relating to performance would be made. Findings may suggest, for example, that technical and demand coefficients at functions or stages one and two were soundly conceived and based, but this was not true at function three. The overall process from function three forward was jeopardized economically because decisions were made using inadequate data or data were not interpreted correctly. A well-trained clinician, moving with stethoscopic and pulse-measuring steps, could spot the problem(s) and suggest remedial steps. Whether the problem area(s) was production or marketing in the traditional sense makes no difference. There was a problem of insufficient, inadequate, or wrongly interpreted data which resulted in combinations of resources which were not economically sound. The solution would be sought within this context.

The umbrella concept permits the inclusion of any and all institutional arrangements within the decision-making framework. These become an integral part of the process. It permits and encourages the suggestion of refinements, changes, or elimination of rules, regulations, or other arrangements which can be shown to be not in the firm's best interest, but also not in the interest of the public. It even permits and encourages the suggestion of new, more appropriate institutional arrangements that take unique characteristics of any part of the decision-making process into account in such a way that the output of goods and/or services and the process of producing them and making them available are more consonant with the public interest. This, too, is a part of the clinical or diagnostic posture encouraged by use of the umbrella concept.

ROLE OF THE MARKETING SYSTEM

The role of the marketing system in the umbrella concept is to provide the intelligence necessary to sound decision making at each decision point under the umbrella. Thus, it is saddled with the tremendously important role of providing the information necessary to all points if the system is to be coordinated in such a way that the combinations and recombinations of resources are to be in accordance with relevant specifications.

It recognizes that firms pursue their own individual goals and that these may or may not be in accord with society's wishes. But the perspective suggested is broader and wider on the part of the diagnostician than is usually assumed. A larger picture must be brought into focus if the system is to be properly perceived with the view to making it work.

This mechanism could be viewed as a form of the Smithian Wand. The automaticity which Smith assumed would be lacking, of course. It could, however, provide the basis for a framework within which structural arrangements and their relevancy with respect to economic activity would be considered. The diagnostic posture which is encouraged might very well lead, in many cases, to very positive solutions to problems. It may even prevent the use of poorly conceived measures which serve only to compound the problem(s). It would encourage a cost-benefit approach to problem diagnosis and solving. It could result in many suggestions that bring about resource combinations and outcomes not only consistent with society's goals, but also with the goals of the firm.

ECONOMIC BASES FOR GROUP ACTION

Definitions of marketing, the marketing umbrella concept, the diagnostic and clinical approach to problem solving, and the role of the marketing system are of interest, per se, hopefully. However, they take on relevance for our purposes only as they help us develop our position regarding the why of agricultural cooperatives. Let's use them as building blocks toward that objective.

STRUCTURE—COMPETITION

Reference has been made a number of times to the competitive structure of the industry as being important in considering questions relating to whether the goals of a firm are consistent with goals of society. It would be logical to assume that society wishes to have an ample supply of pure and wholesome food made available to it, when it is desired, in an efficient and safe manner and at reasonable prices. Can we assume that there is a parallelism between the firm's goals and the goals of society? Does it exist at any or all of the functional areas?

Goals are performance related whether they are societal or those of a firm, and there may or may not be unity between them. Raising questions regarding conformity between the two within the context of competitive structure implies that a relationship between such conformity and structure of the industry may exist. This proposition will now be explored.

Let us begin by making clear what is meant by competitive or competition and by structure. If parallelism of a firm's goals with those of society is related to or dependent upon an industry's competitive structure, it is important that these terms be understood. Let's also make some reference to what is meant when we refer to an industry.

COMPETITION

Competition in the marketplace and the optimal solution to economic problems resulting from the action of competitive forces are often associated with Adam Smith's "invisible hand." To Smith, the basis for the successful functioning of a market economy was the unfettered pursuit of individual self-interest controlled by competition. According to Smith, as each individual pursues his own self-interests and strives to maximize the value of his own capital, he " . . . renders the annual revenue of society as great as he can. He generally, indeed, neither intends to promote the public interest, nor knows how much he is promoting it. He intends only his own gain, and he is in this, as in many other cases, led by an invisible hand to promote an end which was no part of his intention."

The invisible hand in Smith's view was market prices resulting from the interplay of competitive forces. If these forces are not permitted to work, say by monopoly, resources will not be allocated in the interest of society.

The economic model espoused by Adam Smith is known as perfect or pure competition. The model he denounced as a frustrator of the interests of society was that of monopoly. By interfering with the workings of the invisible hand, the resource allocation and income distribution jobs on the part of the monopolist might well lead to bettering the interests of the monopolist at the expense of societal goals. Let's examine these two models.

The competitive model rests on the assumption that there is competition, but what do we mean by competition?

To Smith, competition was an independent striving for sales by the sellers in a market. He pointed out that independent action could emerge where there were only two sellers, but it was very unlikely. He recognized short-run immobility of resources, but if artificial barriers to resource shifts were not put in place, the full benefits of the competitive process would be realized in the long run.

Later, the structural aspect of competition was more strongly developed. Structure refers essentially to the number of buyers or sellers in a market. An industry is said to be competitive only when the number of firms selling a homogeneous product is so large and the share of the market of each individual firm is so small that the market price of the commodity is not affected by the firm's actions in varying the quantity of the commodity it sells.

As shown in Fig. 1.2, price to an individual firm under conditions of pure competition is a given or a constant. It is determined by total market forces and the individual seller has no control over its level or movement. Sales by an individual seller under these conditions could be doubled or cut in half (or withdrawn completely for that matter) and no perceptible influence on price would result. The seller thus is a price taker. There is no ability to affect price in any way. Any actions or output decisions are not taken into account in the planning operations of the other sellers in the market, since there is no perceptible economic impact regardless of what they do as individuals.

As noted, the crucial element involved in the competitive model is the number and size of the sellers in a market. The competition is on the basis of price. Each individual firm in the market can, from an economic standpoint, ignore all other firms because the economic impact of any actions they may take will not be perceptible on the market price.

MONOPOLY

Let's now consider another economic model based upon structural characteristics. This is the model of monopoly which Smith denounced as frustrating the unity between the objectives of the firm and those of

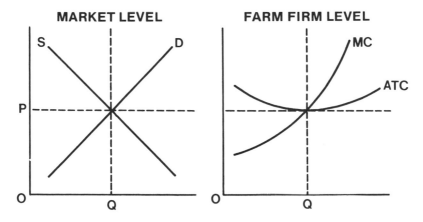

FIG. 1.2. Market price to an individual firm under conditions of pure competition.

society, objectives which he said were consistent with each other under conditions of pure competition.

As the name indicates, monopoly refers to a single seller. Despite the fact that true monopolies exist only under conditions of state sanction, let's consider some of the areas of conduct which are available to firms with this structural feature. This will be helpful in meeting our objectives.

Operating in this structural arrangement opens the door to many areas of conduct not available to firms operating in industries that meet the conditions of the competitive model—mainly a large number of small sellers. Whereas firms in a competitive environment act independently of other firms in the industry, in this case the firm is the industry. With this structural condition many conduct avenues are opened which are not available to firms operating in a competitive environment. Barriers to entry through product differentiation and costing strategy can be erected. Other firms find it difficult if not impossible to enter. Charges of lethargy in the area of research and development and the acceptance of new technology have been leveled at firms with this structural arrangement. Product strategy relating to variety, quantity, and quality are, again, independently determined, but with goals in mind which may or may not be consonant with those of society.

The economies of scale factors and conditions leading to so-called natural monopoly and public regulation, as with utilities and the railroads, will not be discussed here. Our objective is to emphasize that industries that are structurally arranged toward the monopoly end of the spectrum have areas of conduct, especially in pricing their product, open to them which are not open to industries structurally tending toward the competitive model.

THE RANGE OF COMPETITION

As an aid in picturing more clearly in our minds how competition in the marketplace is related to the numbers and size of firms making up an industry, a range of competition is shown in Fig. 1.3.

As noted, agriculture is positioned near the pure competition end of the range. Despite the fact that the number of farms has decreased and the average size of individual farm firms has increased, agriculture still most closely approximates the Smithian concept of competition. The individual firm has no ability to affect price in the marketplace by any action it may take. The individual firm gives no thought as to what other

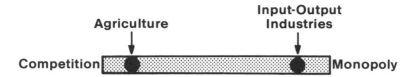

FIG. 1.3. Range of price competition based on numbers and size of firms within an industry.

firms will do if it takes any kind of action such as offering all or any part of its supply of product for sale. This lack of concern and the independent position of each firm reflects their knowledge that the share of the total supply of each firm is so small that it really makes no difference what they do or do not do. This situation is closest to that visualized by Adam Smith.

The monopoly situation is found at the other end of the range. As indicated before, no monopolies exist except those that are publicly sanctioned, such as public utilities. As shown on the scale, however, those firms that buy the output of agriculture and those that provide inputs to agriculture are positioned toward the monopoly end. This means that these industries are made up of a relatively few large firms. Each firm has such a large share of the total market that its actions have an impact upon the marketplace. It has the ability to control its output and thus price. The firms in this competitive category, referred to as monopolistic or monopsonistic competition, are not independent of each other. Under these conditions, each firm considers what action might be taken by the other firms in the industry in response to any action it might take. Such actions might include reducing or increasing output, raising or lowering prices, and launching advertising and promotion campaigns in an effort to increase its share of the market. Each firm has the ability to engage in these tactics, but other firms, because they, too, are large and control some significant share of the market, have the ability to react with similar tactics.

The actual situation which exists, as indicated, is a spectrum of structural configurations along with associated conduct categories from pure competition to monopoly. No industry would meet the structural and other requirements of the competitive model, and none, except for publicly regulated monopolies, would fall into the monopoly category. This means that most industries fall somewhere between pure competition and monopoly. A condition of monopoly shades gradually into one of price competition as the sellers increase in number. Concomitantly, the conduct options open to firms within one industry move in the same direction. Only when sellers are so numerous that each seller has no perceptible share of the market do sellers act independently of one another. Farmers, for example, view each other as neighbors rather than rivals in the economic sense used here.

This means that most of our economy operates in some form of imperfect competition—somewhere between monopoly and pure competition. The reasons for this situation will not be explored here because they are not relevant for our purposes. Suffice it to say that because of scale economies and substitution limitations on the demand side, we find industries made up of large firms with large fixed capital outlays and resultant large overhead costs. Many firms are multiplant. But most importantly for our purposes, this configuration opens up a range of opportunities on the conduct side which, if not addressed by responsible and judicious policy and programs, may leave conduct possible which is not in the public interest.

Let's remember that our immediate objective is to use relevant information in establishing a position we can support in regard to whether agricultural cooperatives are justified. Looking at numbers and sizes of firms within an industry is of interest, per se, but in trying to develop a position with which we feel comfortable, it is necessary for us to go further. The implications of the range of opportunities in the conduct of firms as related to sizes and numbers of firms making up an industry are of greatest importance for helping us to establish a position.

The basic question here, of course, is whether conduct options open to firms as a result of their industry makeup and whether their positions on the range of the competition scale make it potentially possible for the firm's goals to diverge from the goals of the public. In Adam Smith's utopian situation, such a divergence was not possible in the long run. As Smith recognized, however, this is possible where monopoly conditions are involved.

INDUSTRIAL ORGANIZATION MODEL

Figure 1.4 depicts a model of industrial organization which will provide further help in developing our position. It is based upon the premise that a close relationship, almost causal in nature, runs from basic conditions to structure to conduct to performance. The suggestion here, of course, is that certain basic or fundamental conditions on the supply and demand side as associated with particular commodities lead to certain structural arrangements of the firms making up that industry from a numbers–size standpoint. Given the structural arrangement that evolves, conduct options such as pricing behavior, product strategy, and advertising are open to the firms in an industry. Further, given the

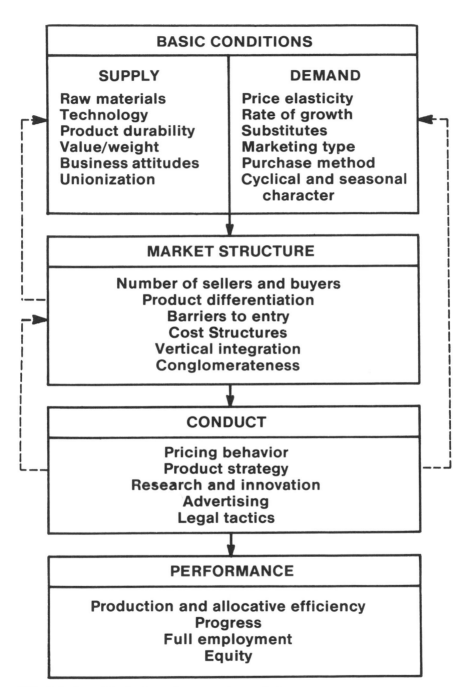

FIG. 1.4. A model of industrial organization analysis.

Source: Scherer, F. M., Industrial Market Structure and Economic Performance, 2nd Edition, p. 4. Copyright © 1980 by Houghton-Mifflin Company, Boston. (Used by permission.)

conduct options that are open to the firms, the outcome or performance of the firms may lead to results which may be in keeping with the firm's objectives, but which may be at variance with those of the public. Criteria for measuring performance are provided in the model and are useful in thinking through questions relating to when and if a firm's goals or objectives, as measured against the suggested criteria, may, if achieved, result in outcomes which are not in keeping with the interest of the public.

This type of analysis is particularly useful for our purpose in trying to reach a well-founded position in regard to whether agricultural cooperatives are justified. Using the basic conditions, market structures, conduct, and performance model with agriculture as the industry being considered leads to certain conclusions with respect to the structure of the industry and what conduct options are open to it. Then something can be said about its performance as measured by the suggested criteria along with whether it is in conformity with the performance considered appropriate by society.

In like manner, other industries can be subjected to the same analytical procedure and conclusions can be reached. In this way, we can move toward our goal of developing a position in regard to whether agricultural cooperatives are justified. This constructively critical posture is a basic component of the process of moving toward professionalism and is a major current objective in our work in this course.

REFERENCES

Breimyer, H. F. 1976. Economics of the Product Markets of Agriculture. Iowa State University Press, Ames.

Goodwin, J. W. 1982. Agricultural Economics, 2nd Edition. Reston Publishing Co., Reston, VA.

Kohls, R. L. 1967. Marketing of Agricultural Products, 3rd Edition. The Macmillan Company, New York.

Rhoades, V. J. 1978. The Agricultural Marketing System. Grid Publishing, Inc., Columbus, OH.

Scherer, F. M. 1980. Industrial Market Structure and Economic Performance, 2nd Edition, p. 4. Houghton-Mifflin Co., Boston.

Shaffer, J. D. 1968. Agricultural organization in the modern industrial economy. *In* The Scientific Industrialization of the U.S. Food and Fiber Sector—Background for Marketing Policy, North Central Regional Publication 20. The Ohio State University, Columbus.

Shepherd, G. S. 1963. Agricultural Price Analysis. Iowa State University Press, Ames.

Smith, Adam. 1937. An inquiry into the causes of The Wealth of Nations, p. 423. Random House, New York.

TO HELP IN LEARNING

1. Restate your objectives in taking this course.

2. Ask several students, business people, farmers, teachers, etc. what their definition of marketing is. Compare theirs with that given in this chapter. At this stage, what is your conclusion?

3. Ask the same people how they define competition. Have they ever used the terms structure, conduct, and performance in any of their discussions?

4. Talk with students and get some idea of what monopoly, in the structure/conduct/performance framework, means to them.

5. Discuss agricultural cooperatives with several of your peers to determine their interest in, awareness of, and knowledge about them.

6. Summarize your findings in the areas above in a short paper called, "Results of Interviews Regarding Marketing, Market Structure, and Agricultural Cooperatives."

DIRECT QUESTIONS

1. Prepare a definition of marketing for use in a 2-minute talk to your class.

2. What is meant by specifications of the potential uses in terms of economic activity? How may resources be wasted in this context?

3. What is pure competition? What is meant by monopoly? What is meant by the perfect market? How is demand influenced in it?

4. What is meant by the Smithian Wand?

5. What is meant by a production–marketing dichotomy? What is its relevance for our objectives in this course?

6. What is the so-called business school approach to marketing? How is demand influenced in this concept of marketing?

7. What is the structural approach to marketing?

8. Describe what the marketing umbrella means to you.

9. What is the role of the marketing system? Compare this with the production–marketing dichotomy position.

10. What does a clinical–diagnostic approach mean to you?

11. What are the goals of a firm? Of society? What is meant by parallelism between them?

12. What is a price taker? What is a price maker, in the context of our work in this class?

13. What is meant by monopolistic competition?

14. How is the structure/conduct/performance model of industrial organization relevant for our purposes?

15. What does the word professionalism mean to you?

TYING-TOGETHER QUESTIONS

1. The first part of this book is aimed at developing the why of agricultural cooperatives. At this stage, how does a definition of marketing, economic activity, industry structure, and the marketing umbrella contribute to this objective?

2. When would you consider a firm to be performing satisfactorily? What performance criteria would you use?

3. Why are some industries made up of a small number of large firms? Name four or five such industries relevant to our objectives.

4. Why is agriculture at the position indicated on the scale of competition? What does this mean for our purposes?

5. At this stage, what can you say about the why of agricultural cooperatives?

2

Structure of Agricultural Input and Output Industries

Our efforts in this chapter will be directed toward determining the position on the spectrum of competition, ranging from pure competition to monopoly, which is occupied by representatives of the industries providing inputs to agriculture and those buying the output from agriculture. The purpose will be to assess the range of economic opportunities afforded them as a result of their position on the spectrum. The ultimate objective, of course, is to be able to say something regarding strategies relating to price and output which are open to the economic actors with which we are concerned. Hopefully, this will provide bases for policy and program implications and possible prescriptive measures that will be in the public interest. The structural format set forth in the preceding chapter will be used in this effort, especially the range of competition scale as shown in Chapter 1 (p. 13) and the structure, conduct, and performance model of industrial organization at the end of the same chapter.

THE INDUSTRY

Before we proceed further, the concept of an industry should be clarified, with some acceptable definition being advanced. This is essential, since we are concerned with relevant industries, their structural configuration, and their opportunity sets as related to their structural makeup. It was mentioned in the previous chapter that the industry concept was relevant for our purposes and that it would be defined.

The major objective and the major problem in defining an industry is making sure that all firms that are competitors are included and all others are excluded. With today's movements toward integration and conglomeration, this is becoming more and more difficult. Industry leaders' sales, in the category being examined, must in some way be separated from their sales in other categories if meaningful assessments are to be made. For example, petroleum products as a fuel and lubricant are extremely important to agricultural producers as a major input. Petroleum companies, however, are involved in other fuel source areas, such as coal, and also in other areas not so closely related to fuel. Somehow, the categories meaningful to agriculture for our purposes must be separated from those that are less meaningful.

The system most often used in categorizing the output of a business enterprise is the Standard Industrial Classification, or SIC, developed through the Census of Manufacturers by the Bureau of the Census. It is built around a system of 7-digit numbers, with finer and finer degrees of classification or categorization being indicated as digits are added. Thus, the first digit might indicate the firm was in manufacturing or mining or agriculture, for example, a very broad category. Additional digits would be added which would bring further refinement in categorizing the firm until it was pinpointed as meaningfully and as clearly as possible. No 7-digit categories are published by the Census of Manufacturers because of the understandable difficulty in being this precise.

Our goal here, as we recall, is to define an industry as meaningfully as possible for our purposes and to determine where it is found on the pure competition–monopoly spectrum. An industry, then, includes those firms that are competitors in a rather well-defined area. Of course, all firms compete with each other in the sense that they strive to make sales in return for income which might be at our disposal. This could not be done with a 7-digit or even a 70-digit classification and for our purposes would not be appropriate. We wish to consider only those firms that compete directly in selling inputs to agriculture and in buying outputs from agriculture. Once we have these, the relevant industries are defined for our purpose.

MARKET SHARES—CONCENTRATION RATIOS

The dimension of market structure with which we are concerned relates to the possession of market power as associated with the firm's position on the competitive scale. The method most often used in measuring such power employs the SIC categories in determining the number of firms that compete directly (make up an industry) and then determines the market concentration ratio which is the percentage of total industry sales (or some other relevant unit for measurement) accounted for by the largest few firms ranked by market shares. Such ratios are computed with 4-firm, 8-firm, 20-firm, and 50-firm concentration ratios. These are the ratios most commonly published. For example, the percentage of total industry sales (or other relevant unit) accounted for by the 4 largest firms, 8 largest, 20 largest, and 50 largest would be provided.

Problems of reporting by business firms, validity and reliability of data, grouping or separation of operations, different reporting procedures, and problems in properly combining data that are available are inherent in the procedure used in industry classification and the construction of concentration ratios. Such problems make it very difficult to determine just who competes with whom for our purposes. These are recognized, but such ratios will be used in our concern with market power on the part of those firms in those industries that are most closely involved with agricultural producers—those who provide them with inputs and those who buy their output.

SUPPLIERS OF INPUTS TO AGRICULTURE

Let's look at the market concentration ratios of a group of industries that are significant suppliers of inputs to agriculture. These are shown in Table 2.1. Note the market shares of the groups of companies and consider the implications. Compare this with the market share of the four largest dairy farms in the United States, for example.

The trends for nonpurchased inputs (those grown on the farm, such as feed for horse and mule power) and for purchased inputs (e.g., petroleum, farm machinery, and chemicals) have moved in opposite directions. As shown in Table 2.2, nonpurchased inputs in 1977 had declined to half those used in 1940. At the same time, purchased inputs more than doubled during this period. This suggests a growing dependency of agriculture upon other sectors.

As discussed, labor use declined rapidly as capital was substituted for

TABLE 2.1

Concentration Ratios for Representative Industries Supplying Inputs to Agriculture, 1977 and 1972, Based upon Shares of Value of Shipments[a]

SIC codes	Class of product	Year	Share of value of shipments (percentage accounted for)			
			4 largest companies	8 largest companies	20 largest companies	50 largest companies
28694	Pesticides (and other inorganic agrichemicals)	1977	65	80	93	99+
		1972	57	79	97	100
2873	Nitrogenous fertilizers	1977	28	45	72	96
		1972	26	43	71	96
2874	Phosphatic fertilizers	1977	49	72	99+	100
		1972	53	80	99+	100
28796	Herbicidal preparations, agricultural, gardens, and health service use	1977	65	84	96	99+
		1972	77	89	98	99+
29111	Gasoline	1977	30	53	84	98
		1972	31	55	86	98
29115	Lubricating oils and grease	1977	45	69	98	100
		1972	39	64	98	100
3011	Tires and inner tubes	1977	71	89	97	99+
		1972	73	89	97	99+

SIC code	Product	Year				
32410	Cement (including cost of shipping containers)	1977	24	41	75	99
		1972	26	47	79	99+
32740	Lime (including cost of shipping containers)	1977	34	49	77	99
		1972	36	52	79	99
33156	Fencing and fence gates	1977	46	72	97	100
		1972	40	64	94	100
34444	Metal roofing and roof drainage equipment	1977	25	39	62	84
		1972	34	46	69	90
35231	Wheel tractors and attachments	1977	80	99	—	—
		1972	81	98	100	—
35233	Planting, seed, and fertilizer machines	1977	50	60	79	94
		1972	43	56	76	93
35235	Harvesting machinery (except hay)	1977	79	87	94	99
		1972	71	81	92	98
35237	Plows	1977	69	78	92	99
		1972	70	81	93	100
37111	Passenger cars	1977	99+	99+	100	—
		1972	99+	—	100	—
37112	Truck tractors, chassis, and trucks	1977	85	98	99+	100
		1972	84	97	99+	100

[a]Data from which concentration ratios are calculated are based on 5-year interval census data. Ratios based on 1982 data will not be available until late 1987.

Source: Concentration Ratios in Manufacturing, MC 77-SR-9, U.S. Department of Commerce, Bureau of the Census, May 1981, applicable pages.

TABLE 2.2
Indexes of Farm Input Purchases, 1940–1977

	Inputs (1967 = 100)					
Year	Non-purchased[a]	Purchased[b]	Farm labor[c]	Mechanical power and machinery[d]	Agricultural chemicals[e]	Feed, seed, and livestock[f]
1940	159	58	293	42	13	42
1950	150	70	217	84	29	63
1960	119	86	145	97	49	84
1965	103	93	110	94	75	93
1970	97	102	89	100	115	104
1975	92	107	76	113	127	101
1977	88	118	71	116	151	110

[a]Includes operator and unpaid family labor, and operator-owned real estate and other capital inputs.
[b]Includes all inputs other than nonpurchased inputs.
[c]Includes hired, operator, and unpaid family labor.
[d]Includes interest and depreciation on mechanical power and machinery, repairs, licenses, and fuel.
[e]Includes fertilizer, lime, and pesticides.
[f]Includes nonfarm value of feed, seed, and livestock purchases.

Source: Hamm, L. G., in Structure Issues of American Agriculture, USDA, ESCS, Agricultural Economics Report 438, November 1979, p. 219.

labor. Machinery and chemicals were major items in this substitution mix. Mechanical power increased dramatically, as shown in Table 2.2, but not as rapidly as did the use of chemicals. The largest increase in purchased inputs was for chemical fertilizers. The use of pesticides and herbicides also increased rapidly, but this occurred more recently. Mechanical energy and labor have almost completely been replaced by reduced tillage, no tillage, and by herbicides. These items, which have been a part of the factor substitution process, require heavy use of increasingly expensive petroleum-based chemicals.

The price trends for purchased inputs have significance for our purposes. In the past 12 years, tractors and fuel prices have more than doubled (see Table 2.3). Prices of agricultural chemicals have risen, but not as much as have those of tractors and fuel. Feed prices are higher, but have not risen as much as the others. In the aggregate, prices that farmers pay for purchased inputs have more than doubled during the past 12 years.

Hamm (1979), in his discussion of trends in farm input industries, indicates that each supplying industry has unique structural and be-

TABLE 2.3
Indexes of Prices Paid by Farmers, 1965–1977 (1967 = 100)

Year	Production items with farm origin	Production items with nonfarm origin	Feed	Fertilizer	Agricultural chemicals	Fuels and energy	Tractors and self-propelled machinery	Interest[a]
1965	94	94	97	103	98	98	92	79
1970	111	113	101	88	98	104	116	134
1975	169	198	187	217	160	177	195	262
1978	205	238	183	180	147	212	259	396

[a]Interest on indebtedness secured by farm real estate.

Source: Hamm, L. G., in Structure Issues of American Agriculture, USDA, ESCS, Agricultural Economics Report 438, November 1979, p. 219.

havioral characteristics. He points out that little information is available on individual industries. A common thread, however, is that the firms in each industry are oligopolies and nonprice competition predominates. Efforts to differentiate their products through advertising and promotion are common to the scene. As shown in Table 2.4, producers' purchases of the top four brands of farm inputs were a rather high percentage of the total in given years. Brand names, of course, are a basic part of product differentiation and nonprice competition associated with oligopolies.

CONCENTRATION RATIOS

Concentration ratios of the input supply industries, as shown in Table 2.1, indicate that their position on the competition–monopoly scale is toward the monopoly side. For example, the four largest firms manufacturing pesticides had 65% of the sales in 1977. The same was true of herbicides, which farmers are using in their factor substitutions, in this case for the labor input. Fertilizer manufacturing is not quite as concentrated as herbicides and pesticides. In the case of phosphatic fertilizers, the largest four firms account for 50% of sales. Concentration is especially high in tires and tubes, wheel tractors, harvesting machinery,

TABLE 2.4
Purchases of Farm Inputs by Selected Producers by Brand Manufacturers, 1973–1978[a]

Top four brands	Producers purchasing inputs in					
	1973	1974	1975	1976	1977	1978
Seed corn	59.5[b]	58.8	59.1	61.6	57.3	55.6
Corn insecticide	N/A	59.7	72.9	83.5	83.4	83.5
Two-wheel drive tractors	72.8	79.9	N/A	77.6	76.3	79.9
Combines	86.2	88.0	N/A	86.8	84.4	88.8

[a]Data presented represent actual purchases of randomly selected samples of producer-subscribers of Feedlot Management, Dairy Herd Management, and Hog Farm Management, published by the Miller Publishing Company, Minneapolis, Minnesota. Nearly all producer-subscribers of these publications are Class 1 farms. Therefore, these data tend to reflect the purchase patterns of only the largest farm firms. Variation between years can be expected with this sampling procedure. These numbers do indicate the relative concentration and trends in concentration for these farm inputs.
[b]N/A = not available. Figures are percentages.
Source: Hamm, L. H., in *Structure Issues of American Agriculture,* USDA, ESCS, *Agricultural Economics Report 438, November 1979, p. 221.*

plows, passenger cars, and trucks. In most of the industries, 100% of the sales are accounted for by the 20 largest firms.

BUYERS OF AGRICULTURAL OUTPUT

Let's move now to the industries that buy the output of agriculture. Our concern here will be largely with the food-manufacturing sector as being closest to agriculture. Of course, this sector will reflect demand from the wholesale and retail food sectors and will have implications for an associated problem of farm firms, that of access to markets, which will be discussed later.

As shown in Table 2.5, these industries will also be positioned toward the monopoly side of the competition–monopoly spectrum. However, only in the cases of the canned baby foods and the cereal breakfast foods industries are the concentration ratios extremely high for the four largest firms. In all cases, however, except in the case of fluid milk, in excess of 50% of the value of shipments is accounted for by the 20 largest firms.

As shown in Table 2.6, 50 firms accounted for almost two-thirds of all food-manufacturing corporation assets in 1978. The top 50 food manufacturers accounted for 75% of all media advertising and 90% of network television advertising.

According to Connor (1985), aggregate concentration of food manufacturers is increasing. From 1963 to 1978, asset concentration increased by more than 50%. If the current trend continues, 50 firms will account for all food-manufacturing assets by the year 2000.

It is true that food manufacturers and retailers have only indirect commercial contact with agricultural producers themselves. The raw output produced on farms usually passes through the hands of one or more brokers or first handlers before being sold to manufacturers or retailers. There is some backward vertical integration through direct ownership of farms and there is vertical integration through production or marketing contracts. The extent of these types of relationships varies from product to product.

As will be discussed in a later chapter, the fact that direct contact with farm firms is seldom made by food manufacturers and retailers does not suggest that they have no impact on the business decisions made by farmers. The impact is made and felt through the demand for farm products. An imbalance of information and an imbalance in the terms of trade is the usual case where there are few buyers, oligopsony, and many sellers. So-called thin markets will prevail in a situation in which only a

TABLE 2.5
Concentration Ratios for Representative Industries Buying Output of Agriculture, 1977 and 1972, Based upon Share of Value of Shipments[a]

SIC codes	Class of product	Year	Share of value of shipments (percentage accounted for)			
			4 largest companies	8 largest companies	20 largest companies	50 largest companies
2011	Meat packing plants	1977	21	33	50	66
		1972	26	39	53	67
2017	Poultry and egg processing	1977	21	34	55	85
		1972	(n/a)			
20210	Creamery butter	1977	30	42	64	84
		1972	37	47	61	78
2022	Cheese, natural and processed	1977	38	50	62	75
		1972	40	51	62	74
2026	Fluid milk	1977	17	26	43	58
		1972	17	26	41	54
20321	Canned baby foods	1977	98	—	100	—
		1972	95	99	99+	100
20332	Canned vegetables	1977	38	49	70	89
		1972	35	46	64	85
2037	Frozen fruits and vegetables	1977	23	41	60	86
		1972	28	42	64	86

Code	Product	Year			
2041	Flour and other grain mill products	1977	33	54	76	91
		1972	32	53	76	91
20413	Corn mill products	1977	62	85	96	—
		1972	60	77	92	97
20430	Cereal breakfast foods	1977	81	95	99	100
		1972	84	96	99	100
20481	Poultry feeds (except ducks and geese)	1977	24	37	58	78
		1972	30	42	59	78
20482	Dairy cattle feed	1977	31	41	62	82
		1972	32	43	62	79
20485	Swine feed	1977	46	61	80	92
		1972	38	53	72	86
20486	Beef cattle feed	1977	35	50	73	90
		1972	27	38	57	75
2074	Cotton seed oil mill products	1977	42	59	81	99
		1972	42	55	74	96
2075	Soybean oil mill products	1977	50	71	93	99+
		1972	52	68	90	99
20992	Chips (potato, corn, etc.)	1977	52	61	76	90
		1972	49	61	76	89

[a]Data from which concentration ratios are calculated are based on 5-year interval census data. Ratios based on 1982 data will not be available until late 1987.

Source: Concentration Ratios in Manufacturing, MC 77-SR-9, U.S. Department of Commerce, Bureau of the Census, May 1981, applicable pages.

TABLE 2.6
Aggregate Asset Concentration among the Largest Food-Manufacturing Firms, 1963–1978[a]

Size (largest)	1963	1969	1974	1978
50	42.0	52.7	56.5	63.7
100	53.5	67.4	68.5	74.4
200	67.9	73.4	76.7	81.1

[a]Figures (percentages) are lower bound estimates made on the assumption that each firm in a size class of a minor industry is of equal size; each concentration ratio is constructed so as to maximize the ratio consistent with this assumption. Data for 1978 supplied by the Financial Statistics Program of the Federal Trade Commission.

Source: *Internal Revenue Service, Source Book of Corporation Income Tax Returns, various years.*

few transactions for which the terms of trade are known are publicly announced.

Even when farm products are not purchased by food manufacturers directly, the fact that there are only a few manufacturers who use the product may almost automatically bring about a small number of first handlers. The broiler industry is an example of this.

Economic theory suggests that an atomistically organized sector (this is close to the competitive side on the competition–monopoly spectrum) which is wedged between two oligopolistic ones will pay monopolistically determined prices for its inputs and receive relatively lower prices for its output. Lanzeillotti (1980) has suggested that leading food processing and agricultural input firms possess considerable market power and are inclined to use such power to manage or administer their market situation.

Moore (1959) has indicated that when there is a high level of buyer concentration in a local market, there is a likelihood of price fixing, price leadership, price discrimination, and other forms of pricing behavior associated with this structural arrangement. Also, Farris (1970) with country grain elevators and others have discovered that conduct options open to industries with high concentration ratios are usually exercised.

In summary, we find that the structural arrangement of the industries which provide inputs to agriculture and of those which purchase its output position them close to the monopoly and monopsony end on the competition–monopoly scale. As has been discussed, this positioning provides opportunity sets with respect to terms of trade which are not open to industries positioned close to the competition end of the scale. We shall now see where agriculture is positioned on the scale.

Let's keep in mind at all times that our immediate objective is to reach a defensible position in regard to whether agricultural cooperatives are justified. We are trying to do this by examining data and information which we feel are relevant for this purpose. The data and information, of and by themselves, should be of interest, but if we stop at that point and don't pursue their relevance and significance for our purpose, the why of agricultural cooperatives as we are attempting to develop it in a professional manner may elude us.

REFERENCES

Bureau of the Census 1981. Concentration Ratios in Manufacturing. U.S. Department of Commerce, Washington, DC.

Connor, J. M., Rogers, R. T., Marion, B. W., and Mueller, W. F. 1985. The Food Manufacturing Industries, Structure and Strategies, Performance and Policies. Lexington Books, D. C. Heath and Co., Lexington, MA/Toronto.

Farris, P. 1970. Truck shipments from Indiana country elevators, 1968–69 marketing year. Indiana Agricultural Experiment Station Research Report 376. Purdue University, Lafayette, IN

Greig, W. S. 1984. Economics and Management of Food Processing. AVI Publishing Co., Westport, CT.

Lanzillotti, R. 1980. The Conglomerate Corporation—An Antitrust Law and Economics Symposium, edited by Roger D. Blair, Sponsored by the Public Policy Research Center and The College of Law, University of Florida, Gainesville.

Moore, J. R. 1959. Market Structure and Competitive Behavior in the Dairy Industry, The Present State of Knowledge. Ph.D. Thesis, University of Wisconsin, Madison.

U.S. Department of Agriculture 1979. Structure Issues of American Agriculture, Economics, Statistics, and Cooperative Services. Agricultural Economic Report 438. Washington, DC.

U.S. Internal Revenue Service, Sources. Book of Corporation Income Tax Returns (applicable years). Washington, DC.

TO HELP IN LEARNING

1. Restate our objectives in this course.

2. Prepare a 5-minute presentation for a bag lunch meeting with your peers on "The Definition and Relevance of Concentration Ratios."

3. If someone says that the 8 largest firms supplying inputs to agriculture have 84% of total sales, what does that mean to you?

4. If the 16 largest buyers of agricultural products buy 50% of the total output of agriculture, what does this mean in relation to our objectives?

5. Years ago when agriculture grew or produced a large share of its own inputs, it was said to be more independent. How was it more independent and why did it change?

DIRECT QUESTIONS

1. What is an industry?

2. What are SICs?

3. If industries are so difficult to define, why bother with trying to define them?

4. What is market power? What is its source?

5. Give an example of how capital was substituted for labor in agriculture.

6. What have been the price trends of purchased inputs for agriculture over the past 12 years?

7. What is meant by product differentiation?

8. What are oligopolies? Oligopsonies?

9. What is a "thin" market?

10. What opportunity sets are available to firms positioned toward the monopoly end of the scale of competition?

11. Which ones are available to those positioned near the other end of the scale?

TYING-TOGETHER QUESTIONS

1. How does the market system concept shown by the marketing umbrella have relevance for our purposes?

2. Ponder the statement, "It makes no difference where an industry is positioned on the competition scale." Comment.

3. Is there a public interest element involved in firm sizes in various industries? What is it? Is it automatically realized by we the people? Why or why not?

4. Agriculture was completely vertically integrated in the beginning. Then it specialized. Then it moved from independence to interdependency. Explain exactly what happened, why, and its implications for everyone involved.

5. What are the "goods and bads" stemming from the situation in the previous question? What is the relevance of this for our objectives in this course?

3

The Structure of Agriculture

As indicated in the previous chapter, the structural arrangement of the industries that provide inputs to agriculture and buy the products produced by agricultural firms tends toward the monopoly position. They are made up of a few large firms. Each has a sizable share of the market, and actions of each have an impact on market price. They do not act independently of each other because they are very much aware of potential actions of other firms in the industry. Those who sell to agriculture are price makers with respect to prices received for their products. The buyers of agricultural products are also price makers with respect to the prices at which they offer to buy products from agricultural firms.

THE STRUCTURE OF AGRICULTURE

Let's now look at the structure of agriculture. Where does agriculture fit into the competition–monopoly spectrum? Is agriculture a price maker or a price taker as related to its position from a structural standpoint? (Examine Fig. 1.2 and review Chapter 1 again.)

The number of farms in the United States has decreased after having reached a peak of 7 million in the mid-1930s. At the present time, there are about 2.3 million (see Fig. 3.1).

Average farm size has increased, ranging from about 210 acres in 1950 to 400 acres in 1978 (Fig. 3.2). As the number of farms increased, most of the land that was released was incorporated into other farms. Some land went out of production, especially in the Northeast and South, while some new land in the Southeast and along the Mississippi River was brought into production. Total land used for crops at present is about the same as in the mid-1930s—370 to 380 million acres.

The increase in average farm size is greater when measured in actual dollar sales. When sales are adjusted for inflation, however, the increase is roughly comparable to the changes in number of acres.

From 1966 to 1978, the number of farms with sales of less than $40,000 (1978 dollars) dropped by 50%. Farms with sales of over $200,000 increased from 0.6% in 1960 to 2.4% in 1978. The sales from this group increased from 17% of all sales to 39% during this period. The 50,000 largest farms received 23% of total farm receipts in 1960, 30% in 1967, and 36% in 1977. These 50,000 farms constituted 2% of the total number of farms in 1977.

FORMS OF BUSINESS ORGANIZATION IN AGRICULTURE

There are three primary forms of business organization found in farming and ranching operations in the United States. These are individual or sole proprietorships, partnerships, and corporations. Break-

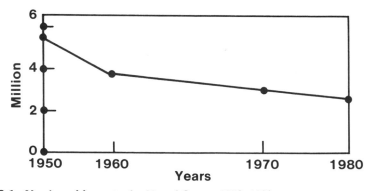

FIG. 3.1. Number of farms in the United States, 1950–1980.

Source: Structure Issues of American Agriculture, Agricultural Economics Report 438, USDA, ESCS, November 1979, p. 25.

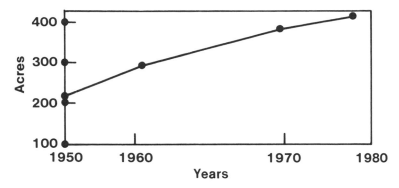

FIG. 3.2. Average farm size, United States, 1950–1980.

Source: Structure Issues of American Agriculture, Agricultural Economics Report 438, USDA, ESCS, November 1979, p. 25.

downs by organizational type, average size, and average annual sales per farm are shown in Table 3.1.

As shown in Table 3.2, individual ownership accounts for about 90% of all farms with sales of over $2500 in 1974. Partnerships made up 9% and corporations 2% of the total. The individual proprietorship type of organization has historically been the main form. Family farm corporations dominate the corporate farm numbers and make use of this form of organization mainly for the purpose of facilitating estate planning and intergenerational transfer of farm assets.

There has been a great deal of attention paid to corporate farming in the past 10 years or so. On a national basis, corporations are most often found in fruits and nuts, vegetables, nursery and forest products, poultry, and cattle production. They account for the sales of more than 25% of each of these products and about 18% of all sales of farm commodities. Sales of all farm corporations in 1974 are shown in Table 3.3

TABLE 3.1
Types of Business Organizations, Agriculture, United States, 1974

Type	Average size (acres)	Average annual sales per farm (dollars)
Individuals	447	36,000
Partnerships	859	77,000
Corporations	3380	417,000

Source: Structure Issues of American Agriculture, USDA, ESCS, Agricultural Economic Report 438, November 1979, p. 30.

TABLE 3.2
Farms, by Form of Organization with Over $2,500 in Annual Sales, United States, 1974

Type	Farms with over $2,500 in sales (thousands)	Number	Acreage (%)	Sales
Individuals	1518	89	75	67
Partnerships	145	9	14	14
Corporations	28	2	11	18
Other	4	a	a	a

[a]Less than 1%.

Source: Structure Issues of American Agriculture, USDA, ESCS, Agricultural Economic Report 438, November 1979, p. 30.

As noted before, corporate farms are relatively large. Each one averaged almost 3400 acres and sold over $400,000 worth of commodities in 1974. As shown in Table 3.3, they vary greatly in size and sales for different types of corporate farms. It is difficult to document the involvement of large, publicly held corporations in agriculture—those that would not be classed as family units. In 1967, it was estimated that perhaps 2500 of the approximately 10,700 Census Class I–IV farm units were in this category. Many of these were in California.

INPUT SUBSTITUTION—GREATER SPECIALIZATION

The amount of farm labor has dropped dramatically in recent years. In 1918, 24 billion hours of labor were used in farm work. By 1950, this had dropped to 15 billion hours. During the mid-1970s, less than 5 billion hours were used per year. Family workers outnumbered hired workers by a ratio of 2 to 1 in 1977.

The decline in farm labor inputs has been offset by increases in use of capital goods such as fertilizer, machinery, fossil fuels, and high-yielding crops and livestock. Fertilizer use increased more than five times since 1950. Use of tractors, as measured by horsepower, has increased almost 150%.

In 1950, labor accounted for almost 40% of the value of all resources used in agriculture. By 1977, this had declined to 14%. In 1950, capital (machinery and chemicals) accounted for 25% of all resources, but by 1977 had increased to 43%. The shift in the resource mix, showing substantial substitution of capital for labor, reflects the changing productivity of inputs and changes in relative prices of the inputs.

TABLE 3.3
Sales of All Farm Corporations, 1974

Items	Share of total U.S. marketings (%)	Distribution of corporation sales among commodities (%)
Grain	5	8
Cotton	16	2
Tobacco	3	[a]
Other field crops[b]	25	10
Vegetables	37	6
Fruits and nuts	32	6
Nursery and forest products	60	7
Poultry	28	12
Dairy	6	4
Cattle	33	41
Other livestock	8	3
All sales	18	100

[a]Less than 1%.
[b]Including peanuts, potatoes, sugar beets, sugarcane, popcorn, and mint.
Source: Structure Issues of American Agriculture, USDA, ESCS, Agricultural Economic Report 438, November 1979, p. 31.

It should also be noted that the substitution process that has taken place reflects a higher and higher degree of specialization in farming. Productivity has increased tremendously, but in the process, agriculture has become more dependent upon purchased inputs and markets for its output and, more importantly, dependent upon the conditions or terms of trade under which inputs are made available and markets are found. Structural characteristics, as reflected by the parties on both sides, again show up as being significant in determination of the final terms of trade.

WHERE IS AGRICULTURE POSITIONED ON THE SCALE?

The picture of agriculture which has been painted shows very clearly an agriculture that from a structural standpoint is much closer to the tenets of the competitive model than are the input suppliers and those who purchase agricultural products. Though trending toward fewer and larger firms for many years, agriculture is still made up of a relatively large number of small farms and is the closest representation of the purely competitive model existing in today's economy. The existence of

large numbers of farms, producing undifferentiated products with relatively easy entry and exit conditions, in an uncertain or risky economic
and production environment makes agriculture the closest example of a
purely competitive industry.

As indicated, agriculture itself has followed a rather typical process of
structural change as a part of economic advancement. This process has
consisted of specialization which led to market exchange and the necessity of considering terms of trade. This in turn led to greater capital
formation, the adoption of new technology, and the substitution of
capital inputs for labor inputs. Increased output per unit of input or
increased productivity has been a by-product. The number of firms has
been reduced, market coordination has increased, and more specialization by firms has resulted. But thousands of farm firms still make up this
industry.

TODAY'S AGRICULTURE

As discussed, as the structure of an industry positions it away from the
pure competition end of the scale, the opportunity sets and strategies
change and give rise to phenomena that are alien to the assumptions of
the competitive model. These include contractual coordination, market
discrimination, nonprice competition, product differentiation, and
bargaining as a replacement for market pricing forces. Agriculture has
taken some of these steps, but a question arises as to whether it will or
can move as individual firms to the situation of monopolistic competition
found with its input suppliers and output buyers.

Indeed, there are those who take the position that agriculture for the
first time in over 40 years is in a relative resource equilibrium position.
They maintain that most of the technical efficiencies available from the
combination of resources for agricultural production can be realized by
farms of a relatively modest size. This position based on the premise that
scale economies are not as available to agriculture as to other industries
suggests once a farm attains a certain size, commodities may not be
produced more cheaply per unit, even if farms become larger.

Use of the competitive model in the problem diagnosis and prescriptive area is criticized by many economists. The argument rests on the
position that it may be rather simplistic in nature in that the problem is
usually diagnosed as a situation in which the industry has deviated from
the competitive ideal, with an implicit solution to the problem—move
back to the competitive ideal. This would suggest policy and programs
designed to fragment those industries that have become concentrated

and to prevent concentration. However, this conclusion fails to take into account legitimate characteristics of an industry which almost dictate that it become concentrated and place itself structurally within the area of monopolistic competition. Policy and programs that recognize this position yet foster desired performance on the part of the industry require great ingenuity on the part of policy formulators.

It seems quite clear from this discussion that agriculture would be placed close to the competitive model position on the competition–monopoly spectrum. Also, it is quite clear that those upon whom the industry is most directly dependent, the input suppliers and output buyers, are aligned much closer to monopoly on this spectrum. As has been stated, this position gives rise to conduct options relating to price and terms of trade which are not open to an industry such as agriculture, an industry that is the closest representative of the competitive model found in today's economy.

It is apparent that when an industry conforming rather closely to the assumptions of the competitive model from a structural standpoint faces input suppliers and output buyers whose structural arrangement opens unique conduct opportunity sets, the competitive firm(s) is at an economic disadvantage. That constitutes a major problem, but there is also another problem which is pretty much dictated by today's very sophisticated and discriminating consumer sector.

PRODUCT SPECIFICATION

A concern with rather specific product requirements such as quantity, quality, timeliness, and uniformity has come to be associated with the options open to the buyers of agricultural products. It has been said that the special characteristic of farming once shaped agricultural marketing institutions, but this situation is rapidly changing. This creates another challenge to agriculture—not to fight specification buying, since its economic underpinning is rather clear. Rather, how can agriculture, with its unique structural arrangement, meet this requirement and at the same time have a voice in setting the terms of trade? In short, how can individual farmers retain or attain access to markets? How can the product offerings of many small farmers be made economically acceptable under conditions of specification buying? Let's explore some of the factors involved in this phenomenon.

Agricultural products have been and are produced on many farms, as was previously shown. A marketing system made up of product assemblers, traders, processors, and wholesalers developed to handle the offerings of thousands of small farms. The system worked, but a system with different requirements has now emerged.

Supermarkets, now the major retail outlet for food, are demanding that specific quantities and qualities of food products be made available to them at a specified time. It has been remarked, for example, that a supermarket chain may very well order X tons of cucumbers, 3 to 4 inches long and 1 inch in diameter, to be unloaded at specific docks between 4:30 and 5:00 o'clock each Thursday morning. In like manner, many fluid milk processors, for example, may order, say, 315,000 pounds of milk, cooled and standardized to a certain butter fat content, and which meets all sanitary and other regulatory requirements, to be pumped into their holding tanks every other day except Saturday and Sunday, at 4:00 a.m. This is known as specification buying and is now commonplace. The question arises as to how the system might be coordinated in such a way that the offerings of many small producers can be assured of meeting the product specifications required. And, of great importance to individual producers, how can they be assured of access to a market? The fact that the impetus for this situation comes from the last function of our marketing umbrella, the consumer through the supermarket, is significant. It is economic reality and must be met in some way.

As farms have evolved from the completely vertically integrated self-sufficient units in the early years to highly specialized firms, the output of each has increased. Farming has become more scientifically based, and the output of products has become more predictable and controllable. But the problem of having a market persists. There are still open markets for some products such as cattle and hogs, but there are no open markets for broilers. Yellow field corn can be sold, but sweet corn and some other types of vegetables can be sold only by those who have contracts.

It is clear that the very process of economic advancement from self-subsistence to market exchange to capital formation to greater efficiency and output serves to highlight the necessity of a system of coordination for the decision makers at the various stages in the process. (Recall the umbrella concept in Chapter 1.) If structural arrangements on the part of some participants are such that the price and production signals of the open market are not working, other methods will have to be considered to bring about the necessary coordination. Contractual arrangements, vertical integration, and perhaps other methods may be considered for use.

The underlying bases for the problem are in the structural configurations that have evolved. This has led to terms-of-trade problems which, in many cases, thwart the transmission of signals for sound decision making. An associated problem relates to specification buying and access to markets by individual producers. These should be kept in mind in seeking a position with which we feel comfortable regarding whether agricultural cooperatives are justified. Such a position, once developed, will serve as the basis for rationalizing public policy and implementary programs that accommodate or omit the use of agricultural cooperatives as vehicles serving the public interest.

STRUCTURE AS IT LEADS TO POLICY

The preceding discussion established the position that the structural arrangement of agriculture places it in the position of being the closest representation of the purely competitive model existing in today's economy. Since it is made up of a large number of farms producing undifferentiated products with entry and exit relatively easy and operating in an uncertain and risky environment, agriculture serves as a prime example of an industry that is closest to being purely competitive. It is positioned close to the competitive position on the competition–monopoly spectrum (see Fig. 1.2).

On the other hand, the position was established that those with whom agriculture is most closely associated and dependent upon—those who provide inputs to it and those who buy its products—are not structurally organized in the same way as is agriculture. These are made up of small numbers of relatively large firms producing differentiated products in an economic environment over which they have some control in regard to supply, demand, and price. They are relatively concentrated industries, as reflected in concentration ratios, with the eight largest firms accounting, in many cases, for large percentages of value of sales (see Tables 2.1 and 2.5). They are positioned close to the monopoly end of the competition–monopoly scale.

It was also pointed out that the goals or objectives of a firm within such industries are not likely to coincide with those of the general public, but that an industry which is structurally arranged, as in agriculture, is much more likely to reflect such conformity. By implication, at least, this means that industries arranged structurally toward the monopoly position will probably have objectives deviating from those of society.

In establishing these positions, recognition was given to the fact that economies of size and perhaps other technical efficiencies may very well push some industries toward the concentrated position. It was suggested that agriculture was not completely within this category and that it may well be that most of the technical efficiencies available to agriculture can be realized by farms of relatively modest size. To the extent this is true, once a farm has reached this size, farm products cannot be produced more cheaply per unit, even if the farm becomes larger. In the dairy farming sector, for example, it is suggested that a unit of about 80 cows with an operator and some hired labor is able to make use of most of the technical capacity available to the sector. As the farm gets larger, if it does, what essentially happens is that a process of adding more units of about 80 cows and complementary technology is followed.

THE FAMILY FARM CONCEPT

This apparent ability to absorb available technology at a relatively small size may relate in a significant way to the public interest and concern over time with the so-called family farm concept. Dating from the Jeffersonian model of what constitutes a family farm to current definitions, this development is of interest for our purposes because of its structural implications.

In the Jeffersonian context, the family farm concept not only made economic sense, but also had political and even moral connotations. Those who held property had a stake in society which they wished to protect, and this made desirable citizens of them. The term family farm was not used in that era, but it was obviously an economic, political, philosophical, and intellectual concept which was a reality. There was no question that this type of structural configuration assured that the goals of the independent, self-supporting landholder were in complete accord with those of society.

Producing for a market was not as alien to the system even at a very early stage as is sometimes taken for granted. Tobacco growers produced for a market in the Jamestown era. Perhaps the rather crude marketing arrangements and the lack of a sophisticated resource allocation process which we associate with modern-day markets serve to perpetuate a notion of complete self-subsistence which we tend to think existed.

The system was destined to become more and more commercial, however, and along with this comes the definitional problem of what constituted a family farm. As the twentieth century came along, it was believed that while a family farm may not necessarily be self-sufficient, as Jefferson contended, it should be able to support a family and keep the family labor fully employed.

The Jeffersonian concept was further modified when it was accepted as a part of the definition of a family farm that some outside labor could be hired. The position that a family farm is one on which the family does most of the farm work with some hired labor was discussed for some time.

Various definitions were considered as the apparent need arose until a definition which is fairly well accepted today was advanced. According to Radoje Nikolitch (1972),

> The essential characteristics of a family farm are not to be found in the kind of tenure or in the size of sales, acreage, or capital investment but in the degree to which productive effort and its reward are vested in the family. The family farm is a primary agricultural business in which the operator is a risk taking manager, who with his family does most of the farm work and performs most of the managerial activities.

The significant feature of this movement of farming from self-subsistence to a commercial, specialized market-oriented farming which characterizes agriculture today is that the structural arrangement has stayed basically the same. Despite getting larger, it is still soundly based from an economic standpoint. Also, it still has support from a political and philosophical standpoint. Perhaps any moral connotations which may have been important considerations in the past have been lost, but the fact remains that the family farm concept persists in basically the same form. It may well get larger as measured by any of the usual units, but it will still find itself relatively close to the competitive pole from a structural standpoint.

AGRICULTURE DISADVANTAGED STRUCTURALLY

This means that agriculture has found itself since the beginning of the industrial revolution at a disadvantage from an economic power standpoint. As it has become more specialized and more market oriented, that disadvantage vis-à-vis input suppliers and output buyers has grown. Specialization and interdependence are two sides of the same coin, as has been pointed out, and this is relevant insofar as the trade concept relates to structural arrangements.

ACCESS TO MARKETS

The disadvantage, as discussed, has persisted over time, but the emergence of the problem of access to markets, discussed previously, is of more recent origin. This is a product of greater and greater market commercialization associated with structural changes which have taken place, mostly in the retailing sector of the food system. Specification buying has come in strong as a means of assuring the quantity, quality, and timeliness of food products made available to consumers.

Thus, the structural disadvantage, on the scene for many years, is joined by the necessity of providing raw materials from the farm which meet specific requirements. The far-flung marketing system that developed to gather the offerings of many small producers is not adequate to meet the requirements of specification buying. The family farm concept persists, however. The goals of the firms in this industry are deemed to mesh more closely with those of society than any other sector. How can structural challenges be met and how can the more current concern regarding specification buying be coped with in such a way that the public interest is best served? Adequate handling of these and similar questions is basic to the development of that position we are pursuing with respect to whether agricultural cooperatives are justified.

ROLE OF GOVERNMENT AND THE PUBLIC INTEREST

Before further consideration of these questions, it is well that we review how government in the United States has chosen to handle areas of conduct of our various economic sectors whose goals may not be in harmony with those of the public. How do we seek to reconcile private interests with the public interest when we cannot depend upon some sort of wand or invisible hand to automatically accomplish this task? A review of economic and legal history as a backdrop for considering this question will be presented. We shall concern ourselves in this review with what might be called the formative period of U.S. antitrust policy. Steps leading to an explicit consideration of agriculture will be examined as they unfold in our economic and legal history. The objective, of course, is to provide a well-rounded account of the development of public policy as it relates to institutions designed to bring public policy and private interests into some sort of acceptable working relationship. This discussion, hopefully, will help in providing better perspective for us in developing the why of agricultural cooperatives. Let's keep this in mind as we examine the succeeding chapters.

REFERENCES

Breimyer, H. F. 1976. Economies of the Product Markets of Agriculture. Iowa State University Press, Ames.

Kyle, L. B., Sundquist, W. B., and Gaither, H. D. 1972. Who Controls Agriculture? *In* Who Will Control Agriculture?—The Trends Underway. North Central Regional Publication *32*. University of Illinois Cooperative Extension Service, Urbana.

Radoje, N. 1972. Family Sized Farms in U.S. Agriculture. U.S. Department of Agriculture, Economic Research Service *499*.

Rohde, G. E. 1980. The Structure of U.S. Agriculture: Some Current and Projected Realities, NICE. Pennsylvania State University.

U.S. Department of Agriculture, ESCS. 1979. Structure Issues of American Agriculture. Agricultural Economics Report *438*.

TO HELP IN LEARNING

1. Agriculture is here on the scale—input suppliers and output buyers are there. What does this all mean for our purposes?
2. Prepare a discussion for one of your classes on agriculture's position on the competition scale and why it is in this position.

3. Conduct a bag lunch seminar on the position on the scale of those industries who buy from agriculture and sell to agriculture and why they are there.

4. Explain to a group of your peers what you believe to be the significance of the situations described above.

5. Ponder the statement, "The public interest may not be automatically served," and comment.

DIRECT QUESTIONS

1. What significance can be attached to the term "share of the market" for our purposes?

2. What is a family farm?

3. Why don't we have, say, 10 to 20 dairy farms in the United States?

4. The labor use, capital substitution process has resulted in what?

5. What are undifferentiated products?

6. What conduct options or opportunity sets are alien to the assumptions of the competitive model? Who has them?

7. What is meant by product specification? Is there any relevance for our objectives?

8. What is meant by market access? Is there any relevance for our goals?

9. Who does specification buying? Why?

10. What is a terms-of-trade problem?

11. What is policy? What is a program?

12. What is meant by the public interest?

TYING-TOGETHER QUESTIONS

1. If pure competition is so desirable, why don't we somehow see that it exists in all industries?

2. What is the significance of the statement that the goals of a firm and those of society may not coincide under some circumstances?

3. What is or should be the basis for public policy and programs to implement it in relation to the previous question?

4. Our objective is to come to a comfortable position in regard to the why of agricultural cooperatives. Up to now, what information, ideas, concepts, etc. have been helpful to you in this process? What are the gaps in the process so far?

5. State your position in regard to the why or justification of agricultural cooperatives as of now and why you hold this position.

4

The Basic Foundation
of U.S. Antitrust Policy

HISTORY'S CONTRIBUTION TO
OUR UNDERSTANDING OF POLICY

As indicated in Chapter 3, it is thought that a review of our economic and legal history as they reflect the environment that spawned our basic antitrust legislation is appropriate. A study of history per se, however, is not the intent at all, although there are those who find such explorations interesting in and of themselves.

In further support of our concern with a lack of understanding of agriculture in general and of agricultural cooperatives in particular, as mentioned in the Preface, we recall a statement attributed to Lincoln, "Cooperatives, let's understand them—a good principle not rightly understood may be worse than a bad principle."

Further, it can be argued that without an acute sense of the past, the present may be meaningless, and the future may not be as productive and fruitful as it might be if it is accompanied by understanding of the evolution of relevant events.

As Sydney Harris, syndicated columnist, wrote (1981), "People die, objects perish, but ideas persist forever. History shows us how ideas have changed people and how people have changed ideas. Lacking this knowledge, we can only be blinded by false passions and betrayed by false hopes."

Two conclusions appear to stem from the above quotations. The first is that there is the strong implication that the principle or fundamental undergirding of the agricultural cooperative as an institutional arrangement may well be good or sound, but if we don't completely understand it, cooperatives may not work. Cooperatives not working in such a case would not be because of weakness in their basic foundations, but because they were not understood. We do not want such a situation to arise in our study of cooperatives.

The second quotation from Mr. Harris makes a case for reviewing the evolution of laws and institutions before we can really understand them. To paraphrase Mr. Harris, how do we know if we want to defend the Capper–Volstead Act (or castigate it) if we do not understand its history and the evolutionary features of its coming into being?

Both conclusions seem to be perfectly in accord with our objectives for the first part of the course—to understand the why of agricultural cooperatives in order to develop a satisfactory position in regard to whether cooperatives are justified. Again, let's keep this objective in mind as we pursue our study and not get bogged down with the feeling that we are studying history for history's sake. We are studying it for a purpose.

THE BASIC FOUNDATION
OF U.S. ANTITRUST POSTURE

Almost the entire edifice of U.S. antitrust legislation rests upon three foundation statutes: (1) the Sherman Act of 1890; (2) the Clayton Act of 1914; and (3) the Federal Trade Commission Act of 1914. Let's examine each of these in an evolutionary context with the view to moving us toward a greater understanding of the why of agricultural cooperatives.

THE SHERMAN ACT OF 1890

The Legal Background

It is generally agreed that the Sherman Act of 1890 can be considered as the principal expression of antitrust policy of the United States. As John Sherman, senator from Ohio, stated:

This bill, as I would have it, has for its single objective to invoke the aid of the courts of the United States to . . . supplement the enforcement of the established rules of the common and statute law by the courts of the several states in dealing with combinations that affect injuriously the industrial liberty of the citizens of these states.

It does not announce a new principle of law, but applies old and well-recognized principles of the common law to the complicated jurisdiction of our state and federal government.

While it is accepted that the legal foundations of the Act are to be found in the common law, Senator Sherman's statement reflects the intense debate that had taken place in Congress for several years prior to the passage of the Act in July 1890. Nevertheless, English common law had its influence, and after national independence the American element increased greatly during the century preceding enactment of the law.

Economic movements involved in and perhaps acting as triggering devices for much of the intense debate that had taken place included the following:

1. Capital-intensive production on a large scale had a rapid rise in the manufacturing industries in the latter half of the nineteenth century. Technological innovation such as use of interchangeable or standardized parts came into being. Transportation costs fell. All movements raised optimum plant scale and set the stage for bigness—industries made up of a small number of large firms.

2. Industrial banking houses were established. Modern capital markets came into being and venture capital became available for the first time.

3. State incorporation laws were liberalized. This contributed to mergers and consolidation and the modern corporation in which stockholders' decision-making power was delegated to full-time managers whose goals and objectives might be different from those of the owners-stockholders.

4. While optimum plant scale from a production standpoint was increasing, so were the markets. Transcontinental railroads created a common market of the United States. Markets were no longer isolated.

5. Two severe depressions also played a part in the way things developed. One in 1873 lasted 6 years and was worldwide in scope. The other in 1883–1886 was not quite as drastic and was not as widespread as the previous one.

It is easy to see that each of the events, singly and jointly, pushed industries into configurations that placed them toward the monopoly end of our scale of competition, as shown in Fig. 1.3. Technological improvements, from a plant production standpoint, and opening of the United States into a sort of common market, all dictated larger plants, larger firms, and greater concentration. The emergence of the banking industry and the availability of risk capital pushed in the same direction. The institutional factor relating to more liberal state incorporation laws made it possible for large corporate structures to emerge in which stockholders leave most of the decision making to professional management. Stock transfers and mergers were made possible.

As mentioned before, the U.S. economy suffered two severe business depressions. This meant that market demand went down and the large-scale operations with huge fixed costs looked for ways to weather the storm. Their first reaction was to cut prices, which would have been a reaction in keeping with the assumptions of the competitive model. This was not viewed with enthusiasm by the managers of the large, capital-intensive, high-fixed-costs plants. The path they chose was one of controlling prices by merging with former competitors and eliminating competition. Mergers took place on an unprecedented scale, with Standard Oil being the pacemaker.

Common law in the United States was against agreements to restrain trade, but it had no impact on the merger movement that took place. Such agreements could be challenged only by the parties to them, and they had no incentive to bring a challenge. Injured private parties could bring a suit, but the odds against their being able to mount a case against the huge firms were great. The pricing practices of the trusts were almost completely beyond the grasp of the law. Monopoly power in a raw form was being used.

These huge trusts were charged by the public with driving thousands of small firms out of business through predatory pricing tactics. Price discrimination was rampant. Fortunes were amassed based upon monopoly power. Farmers in the Midwest joined in the hue and cry as prices for the goods they had to buy rose, while prices for the products they had to sell declined. Wealth was flaunted by those who were wealthy and powerful, and income distribution was becoming more and more unequal. The public began to cry for relief and corrective action.

Several states passed antitrust legislation in the years prior to 1890 and passage of the Sherman Act. At least 14 states had incorporated into their constitutions provisions designed to prevent price fixing and to otherwise restrict competition. At least 13 states had statutory provisions, and some states had both constitutional and statutory provisions.

There is little evidence to indicate that any of the states made determined efforts to enforce their laws. This seeming lack of effort may have been due to several factors. In many cases funds for prevention efforts were lacking. It was difficult to get sufficient data or evidence to secure conviction of alleged offenders, especially a large corporation with almost unlimited resources. An obvious problem was lack of jurisdiction on the part of the states because most cases involved interstate commerce. Courts were reluctant to incur the wrath of corporations because of its potential impact on the state's economy.

Efforts on the part of two states are significant in our concern with factors leading up to our federal antitrust legislation. The New York Senate established a committee in 1888 to investigate the pricing policy of the Sugar Refineries Company. The committee was given a life of only 10 days, but it uncovered a surprising amount of evidence in this period concerning the workings of combinations, not only in sugar, but in other areas as well. There was an obvious lack of enthusiasm on the part of most of the committee because they had not really come to the point of feeling that the dangers said to be associated with combinations were really there. Nevertheless, the committee's work resulted in the New York suit against the Sugar Trust and the Ohio suit against the Standard Oil Trust.

Meanwhile, there was some indication that the national government was beginning to at least recognize that a trust problem did exist. President Cleveland made reference to a possible connection between high tariffs and the rise of trusts in his annual message in 1887. He was much more specific and reflected a much harsher stance regarding trusts in his message the next year in which he said, "Corporations, which should be carefully restrained creatures of the law and the servants of the people, are fast becoming the people's masters." It is interesting to note, however, that legislation aimed at the trust problem which seemed to be recognized was not recommended at that time.

The U.S. House of Representatives passed a resolution in 1888 instructing the Committee on Manufacturers to investigate the trusts. It examined the operations of Standard Oil, sugar, whiskey, and cotton bagging combinations and uncovered much important information. Again, the committee members faced the dilemma of not really being sure in their own minds of a position regarding trusts which they would be willing to defend. They submitted their report without recommendation. The sensational investigation of the House Select Committee of the labor problems in the anthracite coal regions of Pennsylvania in 1887–1888 made very clear that the deliberate abuses relating to production restrictions, price fixing, and treatment of labor were made possible only by the combinations of coal-carrying and mining railroads with the mine operators. This committee stated that these combinations could and should be controlled by legislation.

President Harrison in his first message to Congress in 1889 urged Congress to enact federal legislation when he said, in part, that ". . . they are dangerous conspiracies against the public good and should be made the subject of prohibitory and even penal legislation."

The U.S. Senate appointed a committee in 1889 to investigate the meat produce industry. It was asked to establish the nature of the alleged combinations among the meat packing and dressing companies and the beef- and cattle-carrying railroads. This report, which was unusually readable, furnished conclusive evidence that control of the entire meat products industry was rapidly being concentrated in the hands of the Big Four in Chicago. Their ability to do this had been aided, it was alleged by the committee, by their manipulations of the nation's transportation companies. The committee strongly recommended the passage of an antitrust bill which was then before Congress. This bill, later to be known as the Sherman Act, will be examined in the next section.

Relevance of These Events

The basic relevance of this rather extensive coverage of the events and economic movements leading up to the passage of our basic antitrust legislation becomes more apparent when we refer again to our scale or range of competition diagram depicted in Fig. 1.3 and to the model of industrial organization shown in Fig. 1.4.

As indicated there, as industries shape themselves in terms of numbers and sizes, they position themselves on the scale from the competition end to the monopoly end. Certain types of operations lend themselves to concern with scale economies in which costs per unit of output are lowered. They push themselves toward achieving low per unit costs by increasing plant size. Achieving real scale economies in this manner by taking advantage of available production technologies is in the best interest of the firms and also in the best interest of the public. Operation at the lowest point on the long-run average cost curves achieves this end, which is in the best interest of all.

Implications of Events

In the industrial revolution era, which we've just covered in attempting to understand our antitrust posture in the United States, the series of events all pushed in the direction of firm bigness. The production technology of the particular industries involved, the market expansion made possible by the railroad's contribution to the need for transportation, the banking industry movement with the availability of commercial risk capital, and the liberal incorporation laws all pointed toward bigness. Industries emerged made up of a small number of large firms. They were taking advantage of scale economies by pushing out further and further on the long-run average cost curve. Per unit of output costs are lowered in this manner and, as said before, this is in the interest of the firm and of the public. We can see that their interests are in harmony on the basis of this consideration.

As the plant size grew, however, more capital-intensive operations resulted and more and more of the costs became of a fixed nature. Also, as plant sizes grew larger and such industries became more and more concentrated, they positioned themselves toward the monopoly end of the competition scale. This positioning, as seen in our model of industrial organization in Fig. 1.4, opens up conduct options not available to firms positioned near the competition end. They are able to curtail production and hold prices at current levels or even raise them. In short, if positioned near the monopoly end of the competition scale, they may take actions in accordance with their objectives, and those objectives may not be in harmony with the objectives of society. This is apparently what happened, as described previously, and it brought forth a loud cry from the agricultural sector that it was being forced to pay high prices for its inputs and to take relatively low prices for its products. The philosophical position of our society, which tends to sympathize with the competitive, free market end of the competition scale, was aroused. The stage was set for tangible evidence of that position to be reflected in legislation.

THE SHERMAN ACT COMES FORTH

That expression of a society's sympathy with the competitive, free market end of our scale of competition was destined to carry Senator John Sherman's name, but the literature reveals that he did not actually draft the final version of the bill that became law. It was drafted by Senator George F. Hoar of Massachusetts. The senator from Ohio offered a resolution to the U.S. Senate on July 10, 1888. The motion was unanimously adopted and set the legislative machinery into motion which produced 2 years later the first statute relating to trusts. That resolution, adopted by the Senate, is as follows:

> Resolved, That the Committee on Finance be directed to inquire into and report, in connection with any bill raising or reducing revenue that may be referred to it, such measures as it may deem expedient to set aside, control, restrain, or prohibit all arrangements, contracts, agreements, trusts, or combinations between persons or corporations, made with a view, or which tend to prevent free and full competition in the production, manufacture, or sale of articles of domestic growth or production, or of the sale of articles imported into the United States, or which, against public policy, are designed or tend to foster monopoly or to artificially advance the cost to the consumer of necessary articles of human life, with such penalties and provisions, and as to corporations, with such forfeitures, as will tend to preserve freedom of trade and production, the natural competition of increasing production, the natural competition of increasing production, the lowering of prices by such competition, and the full benefit designed by and hitherto conferred by the policy of the government to protect and encourage American industries by levying duties on imported goods.

It is interesting to note that the resolution relied on the taxing power as the constitutional source of authority for legislation dealing with trust problems. This obviously reflects Sherman's interests, since he was a member of the Committee on Finance for the entire 32 years in which he was a senator and for several periods as its chairman. He was recognized as an expert on public finance and taxation. He was regarded as a conservative and was pro tariff. His position on this, however, was one of moderation, and he had said in discussing tariff proposals that protection should extend only far enough to create competition and not home monopolies. He seemingly did not recognize the inconsistency of his position on the two issues, but this was the case with other politicians as well.

Senator Sherman introduced his first antitrust bill on August 14, 1888. It was aimed at all arrangements, whether contractual or not, that tended to prevent full and free competition in the production, manufacture, or sale of articles imported into the United States and any arrangements that tended to increase the cost to the consumer of any such articles. Double damages were provided for anyone injured by such action, and corporations could lose their corporate franchise. The bill was referred to the Committee on Finance.

Senator Reagan, Democrat of Texas, introduced a bill on the same date. His bill defined a trust as a combination of capital and/or skill by two or more persons to restrict trade, limit or reduce production or increase prices, prevent competition, or create a monopoly. Persons engaged in such trusts would be guilty of a high misdemeanor and subject to fine and imprisonment. Constitutional authority for the bill relied on Congress's power to regulate commerce among states and foreign nations.

It is obvious that the underlying philosophy of the two bills was basically the same. When compared with Reagan's bill, however, Sherman's bill appeared to be amateurish. No reference to constitutional authority was made, and it was not explained as to how the forfeiture of a state corporate franchise could be required by the federal government. The bill was reported on September 11, and the obvious defects in the bill were eliminated.

On January 23, 1889, the Senate again considered trust legislation. Sherman asked his colleagues for a final vote on his bill. The bill was amended in several ways, one of which made the measure dependent upon the commerce power of the Congress rather than the taxing power Sherman thought could be used.

Senator George from Mississippi attacked the bill as being unconstitutional and raised a point which would later prove to be of great interest to farmers. He argued that the bill would cause farmers and laborers to be brought within the punitory provisions of the bill because of innocent and necessary arrangements which they would have made for defensive purposes. Debate on the bill in the Fiftieth Congress closed with Senator George's very effective arguments regarding constitutionality of the proposed legislation, and there were no further proceedings in regard to it.

The first bill introduced in the Fifty-first Congress in the Senate was by Senator Sherman. Two other bills of the same nature were introduced on the same day by Senators George and Reagan.

The George bill is of special interest to agriculture in that he explicitly exempted labor and agricultural organizations from the application of his bill. Also, an undue price enhancement provision was included which would appear again in the farm cooperative legislation that was eventually enacted. The bill was sent to the Committee on Finance and was never reported out. Reagan's bill was sent to the Committee on the Judiciary.

About a dozen bills had been introduced in the House of Representatives during this period, with at least two of them containing provisions similar to those in Reagan's bill, which exempted labor and agriculture from their application.

Meanwhile, Senate debate centered largely on arguments ably advanced by Senator George, who argued that proving intent to prevent free competition beyond a reasonable doubt would be next to impossible. He also argued that the bill was utterly unconstitutional, since it dealt with agreements some of which were completely outside the territorial jurisdiction of Congress—those operating within states. The ultimate impact of Senator George's argument was the general feeling that the Sherman bill would have to be substantially amended before being enacted into law. Despite this, it is significant that a conviction persisted on the part of most of the senators that some legislation regarding trusts should be enacted.

The Finance Committee reported the bill in March 1890. Several amendments had been made, most of which were in response to George's objections. A basic change was that the law would be applicable only to agreements between citizens or corporations of more than one state. Despite the revisions, it was still necessary for the senators to engage in extensive debate. Senator Sherman reflected a bit of impatience when he suggested that no intelligent person could question in good faith the need for legislation of this type. He pointed out that the people of the United States would not endure a king as a political power, and he was sure that they would not endure a king over the production, transportation, and sale of any of the necessities of life. Autocrats of trade with power to prevent competition should not be permitted.

Senator Sherman eloquently defended the bill against the critical argumentation of Senator George. He argued that the bill did not announce a new principle of law, but applied old and well-recognized principles of the common law to the jurisdiction of state and federal government. He pointed out that individual states can and do prevent combinations within the state, but they are limited in their jurisdiction to the state. This bill would enable the courts of the United States to apply the same remedies against trusts which are injurious to the interests of the United States that individual states can apply.

Senator George's criticism dealt at length with proof of intent of a corporation to prevent competition and restrain trade. Senator Sherman argued that in providing a remedy the intention of the combination is immaterial. Intention cannot be proved, but if the effects of the acts of a corporation are injurious, tend to produce evil results, and are against the common good as declared by the law, it may be restrained, punished with a penalty or with damages, and, if the circumstances warranted, it might be deprived of its corporate powers and franchise. He argued, forcefully, that it was the tendency of a corporation and not its intent with which the courts can deal.

His rejoinder to George's attack on the constitutionality of such legislation was also well formulated and presented. The bill, as revised, was directed only against combinations composed of individuals or corporations of different states.

WHERE WAS AGRICULTURE
IN THIS DEBATE?

Other parts of the debate were of interest for our purposes in that they revealed a division of opinion among the senators as to whether farm and labor organizations would be affected by the Sherman bill. Sherman charged the Senate with sidetracking the main question. He argued that his bill did not interfere in the slightest with voluntary associations made to affect public opinion to advance the interests of a particular trade or occupation. It would not, for example, interfere with the Farmers' Alliance because it was an association to advance the interests of the farmers, improve their production methods, and introduce new methods. He argued that no organization could be more beneficial than such associations. He went further and said that they were not business combinations. They were designed to promote their interests and welfare and increase their pay, to get their fair share in the division of production. They would not be affected in the slightest by the bill. Despite his lack of concern that labor and farm organizations might be affected by the bill, Sherman offered an amendment to exempt combinations of laborers made with the view of lessening the number of hours of labor or increasing their wages and concerted price fixing among farmers and horticulturalists. He apparently wished to allay the fears of others. The amendment was accepted without dissent. The Senate passed the bill on April 8, 1890. It was received by the House of Representatives on April 11 and referred to the Judiciary Committee. On April 25, the bill was reported back from the Committee favorably and without amendments. After considerable debate, the House approved the bill with one amendment on May 1, 1890.

The bill with the Bland Amendment was received by the Senate on May 2 and Senator Sherman moved that the Senate concur in the amendment. The bill, however, was referred to the Committee on the Judiciary for consideration. The Bland Amendment relating to contracts and agreements entered into for the purpose of preventing competition in the transportation of persons or property from one state to another evoked a great deal of discussion. The language of the amendment was changed slightly, and the bill was reported back with the new amendment on May 16. A conference between the two Houses was requested and the first Conference Committee met on May 17. Economic philosophies of the conferees were expressed in the discussion and the bill was reported back to the House with the Bland Amendment intact. The next day, the House rejected the conference report and a second conference was arranged. On June 18, the Conference Committee report was presented to the Senate recommending that both Houses recede from their amendments. The report was agreed to without debate or opposition.

The same report was presented to the House on May 20. Brief political discussions were carried on in which both parties tried to give themselves credit for taking the initiative in developing legislation to curb the trusts. After this political jousting, the report was adopted and the bill was passed by a vote of 242 to 0 with 85 members abstaining. It was signed by the Speaker of the House on June 23, by the Vice President on June 24, and on July 2, 1890, President Harrison added his signature to the measure, which later became known as the Sherman Antitrust Act.

AUTHORSHIP INCIDENTAL

The authorship of this fundamental statute is of historical interest, but is only incidental for our purposes. It probably could be rightfully said that every member of the various committees was the author. Legislative interests, sincerity of legislators, philosophic underpinnings of the legislators, and their reflections of the position of their constituents are of interest for our purposes, but again, these are most difficult to measure. Minimum standards of organization and meaningfulness were not reached in these debates. Unsystematic handling and inconsistency in most of the speeches and discussions were the order of every day. One thing that did appear to come forth was that "big business" was represented. This does not mean, however, that "small business" was unrepresented. It is also remembered that the strongest expression of antimonopoly feelings came from western farmers and frustrated Southerners in both Houses. It seems quite clear that a substantial majority of Congress sincerely felt the need for some kind of antitrust legislation, and it may have been too much to expect that straightforward, organized, and consistent discussion devoid of partisanship would lead to quick agreement on legislation. It may well be that the legislation met with the unqualified approval of few, if any, legislators, but that is a basic element in the democratic process. Let's look now to some of the basic provisions of the Act that was finally enacted into law.

BASIC PROVISIONS OF
THE SHERMAN ANTITRUST ACT

The act is divided into eight sections (see Fig. 4.1). The first three define the substantive matter, the offenses, and provide penalties. Section 1 declares "every contract, combination in the form of trust or otherwise, or conspiracy, in restraint of trade or commerce, interstate or foreign to be illegal." Fines of not more than $5000 or by not more than 1 year in prison, or both, at the discretion of the court are provided.

Section 2 declares that every person who shall monopolize or attempt to monopolize or combine or conspire with any other person or persons to monopolize any part of interstate or foreign commerce is guilty of a misdemeanor and subject to the above punishment.

Section 3 makes the provisions of Section 1 applicable to the District of Columbia. This would appear to be irrelevant, but it is interesting that Section 2 was not explicitly made applicable to the District.

Sections 4–8 relate to jurisdiction and procedural matters that are not of relevance in this discussion. A few additional remarks relating to the ideological atmosphere in which the Act was conceived, gestated, and finally born do seem to be in order.

UNDERLYING TONE—TOWARD COMPETITION

There can be no doubt that the philosophy of the leading sponsor of antitrust legislation, Senator Sherman, as a believer in a private enterprise system based on full and free competition, was typical of the vast majority of the congressional members. They felt no need to explore an analysis of underlying economic theory. The norm and desirability of free competition was too self-evident to be debated and too obvious to be asserted. Competition was the normal way of business life. Business operated best when left alone, and the natural role of government was that of a policeman in keeping the road open for everyone. Government should remove obstacles to the free flow of commerce and not become an obstacle itself.

Questions have been raised as to why Congress did not legislate competition in a positive manner rather than legislate against trends. Intent is most difficult to ascertain, but it seems clear that the merits of competition were so clear in the minds of the legislators that they naturally assumed it was more important to get a clear picture of the evil to be prevented than to try to impose positive action patterns on the

CHAP. 647.—An act to protect trade and commerce against unlawful restraints and monopolies.

Be it enacted by the Senate and House of Representatives of the United States of America in Congress assembled,

SEC. 1. Every contract, combination in the form of trust or otherwise, or conspiracy, in restraint of trade or commerce among the several States, or with foreign nations, is hereby declared to be illegal. Every person who shall make any such contract or engage in any such combination or conspiracy, shall be deemed guilty of a misdemeanor, and, on conviction thereof, shall be punished by fine not exceeding five thousand dollars, or by imprisonment not exceeding one year, or by both said punishments, in the discretion of the court.

SEC. 2. Every person who shall monopolize, or attempt to monopolize, or combine or conspire with any other person or persons, to monopolize any part of the trade or commerce among the several States, or with foreign nations, shall be deemed guilty of a misdemeanor, and, on conviction thereof, shall be punished by fine not exceeding five thousand dollars, or by imprisonment not exceeding one year, or by both said punishments, in the discretion of the court.

SEC. 3. Every contract, combination in form of trust or otherwise, or conspiracy, in restraint of trade or commerce in any Territory of the United States or of the District of Columbia, or in restraint of trade or commerce between any such Territory and another, or between any such Territory or Territories and any State or States or the District of Columbia, or with foreign nations, or between the District of Columbia and any State or States or foreign nations, is hereby declared illegal. Every person who shall make any such contract or engage in any such combination or conspiracy, shall be deemed guilty of a misdemeanor, and, on conviction thereof, shall be punished by fine not exceeding five thousand dollars, or by imprisonment not exceeding one year, or by both said punishments, in the discretion of the court.

SEC. 4. The several circuit courts of the United States are hereby invested with jurisdiction to prevent and restrain violations of this act; and it shall be the duty of the several district attorneys of the United States, in their respective districts, under the direction of the Attorney General, to institute proceedings in equity to prevent and restrain such violations. Such proceedings may be by way of petition setting forth the case and praying that such violation shall be enjoined or otherwise prohibited. When the parties complained of shall have been duly notified of such petition the court shall proceed, as soon as may be, to the hearing and determination of the case; and pending such petition and before final decree, the court may at any time make such temporary restraining order or prohibition as shall be deemed just in the premises.

SEC. 5. Whenever it shall appear to the court before which any proceeding under section four of this act may be pending, that the ends of justice require that other parties should be brought before the court, the court may cause them to be summoned, whether they reside in the district in which the court is held or not; and subpoenas to that end may be served in any district by the marshal thereof.

SEC. 6. Any property owned under any contract or by any combination, or pursuant to any conspiracy (and being the subject thereof) mentioned in section one of this act, and being in the course of transportation from one State to another, or to a foreign country, shall be forfeited to the United States, and may be seized and condemned by like proceedings as those provided by law for the forfeiture, seizure, and condemnation of property imported into the United States contrary to law.

SEC. 7. Any person who shall be injured in his business or property by any other person or corporation by reason of anything forbidden or declared to be unlawful by this act, may sue therefor in any circuit court of the United States in the district in which the defendant resides or is found, without respect to the amount in controversy, and shall recover three fold the damages by him sustained, and the costs of suit, including a reasonable attorney's fee.

SEC. 8. That the word "person," or "persons," wherever used in this act shall be deemed to include corporations and associations existing under or authorized by the laws of either the United States, the laws of any of the Territories, the laws of any State, or the laws of any foreign country.

Approved, July 2, 1890.

FIG. 4.1. The Sherman Act, as signed into law by President Harrison on July 2, 1890.

citizenry. Reprehensible behavior was to be prohibited, and monopoly was economically and morally reprehensible.

LABOR AND AGRICULTURE

It is of interest for our purposes that a great deal of literature has been prepared on the relationship of labor unions to the antitrust laws. Interest is enhanced when it is found that, for some reason, farmers' and horticulturalists' organizations for marketing purposes are included in the context of labor unions.

The most controversial single question in the entire field is whether Congress intended to include labor and farm organizations in the coverage of the Sherman Act. It is recalled that Senator George first observed the possibility of conflict between the activities of such organizations and the proposed legislation. He vigorously criticized the fact that labor and farm groups seemed to fall within the framework of the bill. He returned to the subject in his own bill in the Fifty-first Congress which contained an explicit exemption of labor and farm organizations. Of the 18 bills in the House, 8 provided such an exemption by adopting the very wording of George's bill. Not a single bill during the entire period explicitly included these types of organizations within its scope, and this may provide substance to the position that they too were taken for granted as an acceptable way of life, as was competition, and needed no explication.

This problem was brought up by several speakers and Senator Sherman declared that his bill had nothing to do with farm and labor organizations. He went further the following day and proposed an amendment to the pending bill which would incorporate the farm and labor exemption formulated by Senator George. It was adopted without dissent.

Only one member of Congress, Senator Edwards, voiced opposition to the exemption, and Senator Hoar immediately rose to its defense. But the bill as finally reported contained no exemption of labor and farm organizations. The House was completely silent on this point despite the fact that 8 of the 18 bills considered contained such an exclusion.

It is most difficult to interpret such an omission as an expression of a complete reversal of congressional opinion. It seems logical to assume, as alluded to, that Congress had reflected what they considered to be an obvious position that labor and farm organizations were really exempt from antitrust legislation by the very nature of things.

This seems to be a plausible interpretation of congressional action or inaction in this case. The courts had by that time accepted labor unionism as an established social institution and had ceased to regard purely bargaining functions as restraints of trade.

The special status of farm and horticultural marketing organizations may not have been equally well recognized in the courts, but there is little, if any basis for a different interpretation in their case than that of labor organizations. Nevertheless, this omission provided fertile ground for intense and prolonged argumentation before definitive clarification emerged in the form of legislation relating explicitly to farm organizations. The movement toward and the unfolding of this legislation will now be explored.

REFERENCES

Bergh, A. E. (Ed.). 1908. Grover Cleveland Addresses. State Papers and Letters, The Sun Dial Classics Co., New York.

Buck, S. J. 1913. The Granger Movement, A Study of Agricultural Organization and Its Political, Economic, and Social Manifestations, 1870–1880. University of Nebraska Press, Lincoln; Harvard University Press, Cambridge, MA.

Cloud, D. C. 1873. Monopolies and the People, 6th Edition. Day, Egbert, and Fidlar, Davenport, IA; Allen Broomball, Muscatine, IA.

Congressional Record, 7; 6041, 50th Congress, May 1888.

Congressional Record, 3; 2456, 2457, 51st Congress, March 1890.

Harris, S. 1981. Detroit Free Press, Tuesday, October 6, 1981. Used by permission of Mr. Harris.

Letwin, W. 1965. Law and Economic Policy in the United States, The Evolution of the Sherman Antitrust Act. Random House, New York.

Martin, E. W. 1874. History of the Grange Movement, or the Farmer's War against Monopolies. National Publishing Co., Chicago, IL.

McBride, G. 1982. Basics of Cooperatives, the Economic Foundation of Group Action, the History of the Capper–Volstead Act, and the Role of the Member in a Cooperative. 3/4-inch videotape, 49:00 minutes. Michigan State University, East Lansing.

Paxson, F. L. 1924. History of the American Frontier, 1763–1893, Student's Edition. Houghton-Mifflin Co., Boston.

Richardson, J. D. 1897. A Computation of the Messages and Papers of the Presidents. Prepared pursuant to an act of the 52nd Congress of the U.S., Vol. II.

Scherer, F. M. 1980. Industrial Market Structure and Economic Performance, 2nd Edition, p. 4. Houghton-Mifflin Co., Boston. Copyright (c) 1980. (Used by permission).

Schlesinger, A. M. 1927. A History of American Life in Twelve Volumes. The Macmillan Co., New York.

Schriver, H. 1984. The Basics of Cooperatives, NICE. Bozeman, MT.

Thorelli, H. B. 1954. The Federal Antitrust Policy, Origination of an American Tradition. George Allen and Unwin, Ltd., London; P. A. Norstedt and Soner, Stockholm.

Valko, L. 1981. Cooperative Laws in the U.S.A. Bulletin 0902. Washington State University, Pullman.

TO HELP IN LEARNING

1. Review your objectives in this course. Prepare a short statement regarding history and its relevance to your goals.

2. Would you be in a better position to, say, defend the Bill of Rights (or to condemn it) if you understood its historical evolution? Why?

3. Consider this partial statement, finish it, and then discuss it with your peers, "The Sherman Antitrust Act reflects a philosophical position against _____."

4. Take either side of this debate topic and develop arguments in support of it, "The Industrial Revolution taught us that bigness is inevitable and we should let it be."

5. A friend saw your notes taken in this class and asked, "What in the world do a family farm, agricultural cooperatives, and the Sherman Antitrust Act have in common?" What is your reply?

DIRECT QUESTIONS

1. Why did the events of the industrial revolution lead to bigness on the part of firms?

2. Why couldn't states regulate industrial power of large firms?

3. Who is the author of the Sherman Antitrust Act?

4. What is meant by, per se, illegal?

5. How many parts or sections are in the Sherman Antitrust Act?

6. Did the Sherman Act refer in any way to agriculture?

TYING-TOGETHER QUESTIONS

1. How and why does the industrial revolution have relevance in understanding the Sherman Act?

2. Why do you suppose agriculture was not explicitly made exempt from the provisions of the Sherman Antitrust Act?

3. Congressional debate on passage of antitrust legislation indicated that such legislation would not interfere with agricultural organizations because they were not business combinations. Comment on this.

4. Why didn't Congress legislate competition positively rather than legislate negatively against large organizations?

5. At this stage, would you say the Sherman Act will serve the purpose for which it was intended? Why or why not?

5

The Aftermath of the Sherman Act and on the Road to Agriculture's Magna Carta

In considering whether the Sherman Act served the purpose for which it was intended, especially from the standpoint of agriculture, let's review what the Act does. As seen in Fig. 4.1, it declared that

> Every contract, combination in the form of trust or otherwise, or conspiracy, in restraint of trade or commerce among the several States, or with foreign nations, is hereby declared to be illegal.

Further, in Section 2 of the Act, it stated

> Every person who shall monopolize, or attempt to monopolize, or combine or conspire with any other person or persons, to monopolize any part of the trade or commerce among the several States, or with foreign nations, shall be deemed guilty of a misdemeanor, and, on conviction thereof, shall be punished by fine not exceeding five thousand dollars, or by imprisonment not exceeding one year, or by both said punishments, in the discretion of the court.

Clearly, Congress implicitly favored an economic environment within which firms would position themselves toward the competitive end of the scale of competition. In expressing concern with firms forming trusts, engaging in conspiracies, forming combinations, or making any attempt to monopolize trade or commerce, they apparently took the position that industries would naturally be made up of large numbers of relatively small firms with little, if any, economic power as individual firms. Economic action of such firms would be designed to serve their interests, but in the process, the public interest would also be served.

If we are justified in following this line of reasoning, we may have an explanation for the fact that no mention was made of agriculture in the Act. Agriculture was already positioned toward the competition end of the scale, they may have reasoned, so they are as they should be. The task was to prevent other industries from becoming organized in such a way that they could monopolize or even attempt to monopolize economic activity in such a way that other industries could be harmed. Any such monopolization or attempt to monopolize was declared illegal, and that should take care of the situation, or so they might have reasoned.

WHAT WAS EXPECTED?

As indicated, there was a rather general feeling on the part of almost everyone, including farmers, that control of the trusts that were getting a grip on the economic life of the nation would remedy the problem of economic power being exercised by them. As indicated before, the Sherman Antitrust Act forbade all combinations in restraint of foreign and interstate trade. The implication was that the industrial associations would be broken into competing parts and price competition would prevail. Economic positions would be determined by technical efficiency and the production of products which satisfied demanding consumers.

WHAT HAPPENED?

The Sherman Act, despite all expectations, proved to be no barrier at all to the growth of powerful business corporations. Collusion and conspiracy were prohibited by the Act, but legal collusion was possible by corporations following the merger route. Mergers, acquisitions, and other legal transformations through which two or more formerly independent firms came under common control were used to gain market dominance.

Standard Oil Company, United States Steel Corporation, General Electric, American Tobacco, du Pont, National Lead, U.S. Rubber,

United Shoe Machinery, Pittsburgh Plate Glass, International Paper, United Fruit, Standard Sanitary, Allis Chalmers, Eastman Kodak, International Salt, International Harvester, and U.S. Gypsum were among the corporations shaped by the merger and consolidation development. Market concentration grew and along with it came market power. Scale economies and efficiencies were involved, but a desire to escape the rigors and price competition of the open market beyond doubt led corporations to seek and follow the merger route.

Whatever the reasons for the mergers, it was obvious that market concentration came about and more and more market power resulted for the firms. It was also obvious that the Sherman Antitrust Act was ineffective in dealing with the problem of trusts, since it did not inhibit the conspiracy and collusion it was designed to prevent. In short, it did not harmonize the profit-seeking behavior of private enterprises with the interests of the public. The law was strictly construed so that the technique of restraint of trade was considered more important than monopoly power itself. Under the "rule of reason" interpretation as set forth by the Supreme Court in 1911, large business combinations were not restricted so long as their methods were not deemed "unreasonable."

IMPACT ON COOPERATIVES

Despite the fact that the development of cooperative marketing associations was in its infancy when the Sherman Antitrust Act was being debated and finally passed, there was some concern that the Act would inhibit farmers in developing cooperative organizations for effective commercial cooperation. Senator Stewart from Nevada, for example, said, "This measure strikes . . . at the very root of cooperation . . . . When capital is combined and strong, it will for a time produce evils, but if you take away the right of cooperation, you take away the power to redress those evils: It gives rise to monopolies that are protected by law against which the people cannot combine." Senator Sherman accepted this argument and proposed an amendment that provided the Act should not be construed to prohibit any arrangements, agreements, associations, or combinations among persons engaged in horticulture or agriculture made with the view to enhancing the price of their own agricultural or horticultural products. This proposed amendment was considered unnecessary and was not included in the final version. This was the only time during the debate that farmer cooperative associations were mentioned.

The statement of Senator Stewart was significant for at least two reasons. It implied that he did not really expect the Sherman Antitrust Act to prevent the formation of trusts and conspiratorial business arrangements. It also implied that he recognized the possible evils of concentration and economic power on the part of industries providing inputs to agriculture and buying the outputs of agriculture; therefore if trusts were not to be prevented and/or their market power not constrained, a countervailing power arrangement on the part of farmers should not be prohibited.

DANGEROUS TO ENGAGE IN CONJECTURE

Any attempt to portray the ideological atmosphere in which this legislation was born and infer the intent of Congress from what was said or not said is fraught with danger. Congressional debate has been accepted by the courts as evidence of legislative intent in some cases, but has been explicitly ruled out in others. Despite this, it is interesting and relevant for our purposes to recall some of the debate and speech language as they reflect a general interest in the problems. More specifically, the question continued to arise regarding whether it was the intent of Congress to include within the scope of the Act the organizations of farmers most relevant for our purposes.

Based upon the speeches and debates that finally led to the passage of the Sherman Act, there can be no doubt that Senator Sherman and the vast majority of the legislators were sincere proponents of a private enterprise system based on the principle of full and free competition. Despite this, or perhaps because of this, they felt little need to develop the underlying economic theory or to engage in explicit extended debate in support of their position. They obviously took the position that the merits and the norm of free competition were too obvious to be debated or even asserted. Competition was the lifeblood of economic activity, and the proper role of government was to remove obstacles to the free flow of commerce rather than become an obstacle itself.

The legislators, without doubt, felt that the ultimate beneficiary in this process was the consumer. This came about because competition would result in increases in meaningful production at progressively lower prices. There was danger from the strange, huge, ruthless, and awe-inspiring elements of business which had been emerging.

Congress was intent on stressing the evil to be remedied and overcome rather than the merits of competition, which were self-evident. The obstacles to free trade were to be eliminated. Trusts, combinations, monopolies, and combinations in restraint of trade as well as those institutions and societal elements that had been responsible for their growth and development were to be abolished. There is no doubt that the ideological position of Congress as reflected in speeches and in the Sherman Act was a projection of the philosophy of competition. Remove monopolistic elements in an industry and full and free competition would automatically emerge. It was not recognized until later that a policy of legislating against monopolies and restraint of trade may not necessarily be the same as enforcing or maintaining free competition.

WAS AGRICULTURE EXEMPT FROM THE ACT?

The most controversial single question in the debate and discussions leading to the passage of the Sherman Act was whether Congress intended to include labor and farm organizations in its coverage.

Senator George, from Mississippi, was the first to note the possibility of conflict between the activities of farm organizations and the proposed legislation. He vigorously criticized the fact that, in his opinion, labor and farm groups seemed to fall within the frameworks of an early version of the Sherman bill. He prepared his own antitrust bill in the Fifty-first Congress and included a clause that explicitly exempted labor and farm organizations. At least 8 of the 18 bills in the House of Representatives provided for such an exemption, in most cases adopting the wording of the George bill. Not a single bill during the entire period expressly included these organizations.

Senator Sherman, taking note of the concerns expressed by the senators, declared that the bill as reported by his Finance Committee had nothing to do with farm and labor organizations. He then proposed an amendment to the pending bill which would exempt farmers and laborers from its provisions. The amendment was adopted unanimously. The bill, as reported to the Senate from the Committee on Judiciary, however, contained no exemption of farm and labor organizations. Whether this represents a complete reversal of congressional opinion or that Congress had indeed accepted Sherman's view that by their very nature farm and labor organizations were really exempted from antitrust legislation is not clear.

The Sherman Act provides only a general indication of the interest of the Congress and leaves its practical interpretation up to the courts. Such interpretations may be different in different courts and may vary from time to time. In fact, they may contradict at times the intent and desires of those who wrote the law.

Despite the fact that language relating to farm groups was not included in the Sherman Act and whether or not the lawmakers thought that farmers were exempt from its provisions by the very nature of things, the Act had some influence on several cases against cooperatives which were brought before the courts. During the two decades following 1890, directors or officers of agricultural marketing cooperatives were indicted in five states under state antitrust laws. One such indictment in Louisiana was brought under the Sherman Act.

Several states took steps to exempt farmer cooperatives from the provisions of the Act. Illinois passed a law in 1893 that declared the provisions of the Act shall not apply to agricultural products while they are in the hands of producers. This provision of the Illinois law was struck down, however, by the U.S. Supreme Court. Similar activities took place in several other states.

As farmer cooperatives became stronger, more numerous, and more able to have something to say about their terms of trade, several attempts were made to outlaw them as being combinations in restraint of trade. In an Iowa case, a hog marketing cooperative was enjoined through the injunctive process from selling and shipping their hogs cooperatively.

The Iowa Supreme Court ruled that the activities of the marketing cooperative in attempting to have something to say about what they received for their hogs were a combination of restraint of trade and a violation of the Sherman Act. This decision reflected a lack of understanding of agriculture and the role of cooperatives. It completely denied the right of farmers to organize into cooperative associations to market their products.

THE FEDERAL TRADE COMMISSION (FTC, 1914)

Congress passed the Federal Trade Commission Act in 1914. It constitutes a part of our basic antitrust posture, but has no direct relevance for agriculture or for agricultural cooperatives. It serves to strengthen the Sherman Act and concerns itself mainly with unfair trade practices.

It has a distinctive feature, rarely found in the area of U.S. jurisprudence. It is permitted to perform both investigatory and adjudicative functions and would appear to violate our separation of powers doctrine. It came into being in recognition of the need for special competence in business affairs. It therefore has specialists in finance, law, economics, and other areas where competencies are needed in investigatory pursuits and the preparation of cases for prosecution. It outlaws unfair methods of competition and leaves it to its five full-time commissioners to determine what practices are unfair. As indicated, it serves as a strengthening element for the Sherman Act, but has little direct relationship to agriculture or to agricultural cooperatives.

THE CLAYTON ACT—1914

The Clayton Act also served to strengthen the Sherman Act. It is known as an incipiency measure in that it permitted steps to be taken against firms before they became full-blown trusts. It is sometimes referred to as the nip-in-the-bud Act.

Regardless of the intent, the result of the Sherman Act's silence regarding the rights of farmers was to leave an area of great uncertainty as to what farmers could or could not do. The per se illegal interpretation of the combination and restraint of trade provisions of the bill weakened the position of agriculture. Even after the rule of reason era of interpretation emerged, the position of cooperative marketing of products by farmers was, to say the least, unclear.

Agricultural leaders asked their representatives in Congress to clear up the situation so that farmers would know just where they stood in relation to the Act. They felt that additional legislation was needed at the federal level. A step in this direction was taken in 1914 when President Wilson signed the Clayton Act into law. This law was designed to supplement existing laws against unlawful restraints and monopolies.

The Clayton bill was introduced in the House by Representative Clayton from Alabama. He resigned from the House soon after the bill was introduced and a representative from Minnesota, Andrew J. Volstead, took over the floor leadership. Mr. Volstead would later become a sponsor and promoter of cooperative legislation that emerged 8 years later.

The Clayton Act received widespread support, especially from labor unions and farm organizations. Such groups had grown rapidly and their influence in Congress had increased. It contained 26 sections, and one of these, Section 6, addressed the unique position of individual farmers and laborers in their efforts to say something about the terms of trade for their products in the marketplace.

Section 6 of the Clayton Act is as follows:

That the labor of a human being is not a commodity or article of commerce. Nothing contained in the antitrust laws shall be construed to forbid the existence and operation of labor, agricultural or horticultural organizations instituted for the purposes of mutual help and not having capital stock or conducted for profit, or to forbid or restrain individual members of such organizations from lawfully carrying

out the legitimate objects thereof; nor shall such organizations of the members
thereof be construed to be illegal combinations or conspiracies in restraint of trade,
under the antitrust laws.

AN IMPROVEMENT, BUT NOT ADEQUATE

The Clayton Act was a positive movement toward clarifying the position of farmers in marketing their products, but it was not adequate.
The nonstock requirement was a complicating factor. Many cooperatives
had capital stock, so their position was left unclear. It certainly helped in
forestalling some of the erroneous interpretations made by the courts in
regard to cooperatives, but it did not provide a special law for them. It
was a supplement rather than an amendment to the Sherman Act.

So again, a clamor arose for legislation that would completely and
adequately clarify the legal status of farm cooperatives. First steps
toward a comprehensive cooperative bill that would accomplish this
purpose were taken in 1917. It would still be a number of years, however, before the legislation, in final form, would emerge.

THE CAPPER–VOLSTEAD ACT

Marketing cooperatives were being perceived more and more by
farmers as a means of partially offsetting their lack of bargaining power
in the marketplace as individuals. The dairy subsector of agriculture was
leading the way in expanding the number of bargaining associations.

Although Section 6 of the Clayton Act was an improvement over the
"per se illegal" and even the rule of reason interpretational positions
taken by the courts in considering restraint of trade and combinations
under the Sherman Act, it was considered an empty victory by many. It
did establish the legality of cooperative organizations if they were established for the purpose of mutual help, without capital stock, and if
they were not conducted for profit. Any deviation from these requirements, however, left an organization or its members subject to penalties
under the Sherman Act and/or the Clayton Act. It was clearly unsatisfactory because of its definitional shortcomings and because it
afforded no protection at all in regard to the practices normally followed
by cooperative associations.

Aided by what appeared to be a feeling on the part of the public that
farmers' organizations were unique because of the structure and other

characteristics of agriculture and should perhaps have treatment that reflected such a position, cooperative leaders began vigorous efforts to secure legislation that would completely remedy the shortcomings of the Clayton Act. A specific and unequivocal statute was needed. It should cover the entire spectrum of cooperative associations, with special reference to the question of restraint of trade.

Since the number of dairy bargaining and marketing associations had been expanding at a rapid rate, they were seemingly most interested in and in need of such a definite statute. The National Milk Producers Federation, formed in 1916, joined with the National Grange, the National Farmers Union, and other farm organizations in forming the National Board of Farm Organizations, with offices in Washington, D.C. In 1917, a resolution was passed by the Board that Congress be urged to adopt legislation that would protect the rights of farmers to "organize and operate cooperative associations without conflict with the antitrust laws." A committee was formed by the Board and instructed to prepare a bill and seek its passage. A bill was prepared and introduced in early 1919 and was referred to the Committee on the Judiciary by both the Senate and House.

The bill, known as the Capper–Hershman Bill, granted greater exemption from the antitrust laws than was provided by the bill finally passed. Despite the fact that it was endorsed by most of the farm organizations, including the American Farm Bureau Federation, after it was formed the committee would not report the bill.

The promoters of the legislation did not give up hope. After the general election in 1920, they renewed their efforts in the new Congress and with a new president, Warren G. Harding. Congressman Volstead from Minnesota, along with Senator Capper from Kansas and Congressman Hershman from California, submitted a revised bill to Congress. Volstead suggested a different strategy for the passage of the bill than had been used with the previous bill. He suggested that the bill should list the powers and rights it granted to farmers and no mention would be made of any prior laws. It would thus not become a mere amendment to either the Clayton or Sherman Acts.

The bill passed by a large majority in the House, but strong resistance appeared in the Senate. A Senate Judiciary Committee, chaired by Senator Walsh from Montana, made major revisions in the bill. Vehement debates were held over the bill in the Senate and it was attacked as "most vicious class legislation." The revised bill was sent back to the House and was dead for that session.

The bill was reintroduced in the Sixty-seventh Congress and the same pattern was repeated—the House passed it but it was drastically revised by the Senate Judiciary Committee, which Senator Walsh chaired. The bill did not pass. The promoters of the legislation, however, had now gained nationwide support and organized themselves for a final push toward passage of the bill.

Congressman Hershman pointed out that Congress had tried to help the farmers solve their problems in the Clayton Act, but had inadvertently used the words relating to nonstock organizations and that it was as necessary for the farmer cooperatives to have capital stock as for any other business. No business can operate without capital stock, and this unintentionally tied the hands of farmers. The new legislation was necessary if farm cooperatives were to be used.

John D. Miller of the Dairymen's League had chaired a League Committee and had written the original proposal. He prepared briefs which he sent to each senator, pointing out that the House version of the Capper–Volstead bill should be enacted because of the ambiguity of Section 6 of the Clayton Act. He suggested strongly that the Senate amendment would be similarly ambiguous and that "no one could safely predict the legal status of farmer cooperatives" if that amendment remained in the legislation. He further suggested that it is like saying to the farmers of the country that we believe you should organize, but you must do nothing after you have organized. The National Board of Farm Organizations sent out a statement made by Congressman Volstead which was critical of the stand taken by Senator Walsh and his committee.

While the Senate was malingering, Secretary of Agriculture Henry C. Wallace got into the act. He arranged a National Agriculture Conference in Washington, D.C. on January 23–27, 1922. The conference was attended by representatives of agriculture, business, labor, and other organizations. President Harding opened the conference and demanded the right of farmers to cooperate in marketing be legally recognized.

The pressure became too great for the Senate and it responded quickly. The bill was passed on February 9, 1922 by a 58 to 1 vote, with 37 abstentions. Even Senator Walsh, who had been a stumbling block to passage of the legislation, voted for it. Section 1 of the bill was amended to require that cooperatives not deal in products of nonmembers more than with members. This amendment was accepted by the House and the bill was sent to the White House. President Harding, who had favored such legislation, signed the bill into law on February 22, 1922. This was hailed by one newspaper as one of the greatest victories ever won by farmers.

AGRICULTURE'S MAGNA CARTA

The Capper–Volstead Act is a very brief law having only two sections. It contrasts in this regard with cooperative legislation in many other countries of the world in which very precise and extended details are included. The law applied to agricultural cooperatives only, with other types of cooperatives having their own legislation.

Section 1 of the Act sets forth clearly that producers of agricultural products may act together in associations, corporate or otherwise, and with or without capital stock in processing, preparing for market, handling, and marketing in interstate commerce the products they produce without being in violation of the Sherman Act or the Clayton Act. They may have marketing agencies in common, and may make agreements and contracts necessary to carry out their purposes.

There were provisions, however, designed to emphasize the position that agriculture is justified in being given special consideration on the basis of having unique characteristics and that producers should be permitted to provide services cooperatively to themselves. These include the following:

> First—that no member of the association is allowed more than one vote because of the amount of stock or membership capital he may own therein, or Second—that the association does not pay dividends on stock or membership capital in excess of 8 per centum per annum. And in any case to the following: Third—that the association shall not deal in the products of nonmembers to an amount greater in value than such as are handled by it for members.

The first two alternative requirements under the Act are in conformity with the Rochdale Principles having to do with democratic control and limited return on capital. They also reflect the service aspect of cooperatives in that they have a character different from that of the ordinary commercial corporation. The motivating force is not control through stock or capital ownership or returns on the investment in stocks. The associations are formed for the purpose of providing a service or services which would or could not be made available under other forms of corporate structure.

The third requirement limiting the value of business that can be done with nonmembers by the association also serves to differentiate the cooperative form of corporation from the ordinary commercial corporation. This again reflects the service orientation of a member-owned and controlled institutional arrangement. Details regarding whether the percentage of nonmembers' business should be at some other, probably lower, level may be argued, but the concept as a differentiating feature is clear.

Another feature of the Act which serves to differentiate it from the usual commercial corporation is embodied in Section 2. This section removes associations of such persons from the original jurisdiction of the Department of Justice and places them under the surveillance of the Department of Agriculture. Immunity from review and regulation by the courts is not conferred upon cooperatives. Initial determination, however, of whether their activities may be tending to run counter to the public interest is put into the hands of a technically informed and perhaps a naturally sympathetic agency.

It was this provision in the Act that sparked much of the debate in the Senate Judiciary Subcommittee. A recommendation was made by that group that the whole section be struck and that a new clause placing the supervising authority in the hands of the Federal Trade Commission be substituted. The House version of the bill was accepted, however, with the supervisory role remaining with the Secretary of Agriculture.

A prominent cooperative leader pointed out that this was, in fact, a two-way responsibility given to the Secretary. He noted that the Secretary of Agriculture can protect cooperative associations from unwarranted prosecution on the charge of unduly enhancing prices but, at the same time, he is saddled with the obligation to protect the public interest. He also pointed out that there is nothing in the Act which prevents the Federal Trade Commission from taking note of any actions on the part of cooperatives, thus serving as a sort of watchdog.

THE ACT'S SIGNIFICANCE

With the enactment of the Capper–Volstead Act into law, it became clear that the position of agricultural cooperatives as a legitimate form of business organization had been established under the law. This form of corporation was brought to stand on the same footing as other businesses. Practices permissible to other corporate forms of business were permissible to the cooperative corporation. Those that were not permissible to other forms were not permissible to the cooperative.

As Judge Lyman S. Hulbert pointed out, agricultural cooperative associations are not exempt from the antitrust laws. The Department of Justice is still entrusted with enforcement of the Sherman Antitrust Act, the Clayton Act, and the Federal Trade Commission Act as they apply to business enterprises in all forms, including the cooperative corporation. In addition, special powers of original jurisdiction have been conferred upon the Department of Agriculture, but without removing cooperatives from the authority of the Justice Department and the Federal Trade Commission. In a sense, an additional level of supervisory authority has been placed over the agricultural cooperative. This should provide for additional scrutiny in making sure it operates in the best interest of the public rather than prompting the false interpretation that it is exempt from the antitrust laws.

A SLOW, TEDIOUS ROUTE, BUT WE FINALLY ARRIVED

This chapter, along with Chapter 4, has attempted to depict the evolution of our country's antitrust position as reflected in our policy and programs relating to economic organization. Although admittedly difficult, an attempt was made to capture the most relevant aspects of the economic and political environments within which antitrust policy was conceived and brought to term. Debate, argument, language, terminology, and perhaps posturing on the part of legislators and others were used as indicators of philosophical positions and of the general economic environment as reflected in the legislation that was passed.

OUR ANTITRUST BASICS

What finally emerged might be depicted as shown in Fig. 5.1.

The Sherman Act itself is accepted as the foundation program of our policy. The Clayton Act may be considered either as an amendment to the Sherman Act or, as some have suggested, a supplement to it. In either case, it moved in the direction of meeting what was seen as a need almost immediately after passage of the Sherman Act—that of clarifying the position of agriculture with respect to the combinations and restraint of trade provisions of the Act. The Federal Trade Commission Act, with its trade practices, regulatory concerns, and combination of investigatory and adjudicative powers, unusual in our legislative, executive, and judicial watchdog constitutional arrangement, is also strongly reflective of our trust or antitrust posture. Since it was not directly involved in our concern with antitrust policy and programs vis-à-vis agriculture, it has not been explored as were the Sherman and Clayton Acts.

Finally, the Capper–Volstead Act evolved, viewed by some as an amendment to the Clayton Act and by others as an extension of that Act

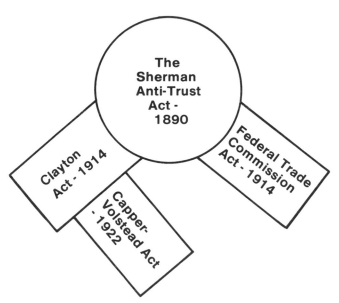

FIG. 5.1. The basic expression of antitrust policy in the United States.

or perhaps of the Sherman Act. It clarified the positions of agriculture with respect to U.S. antitrust policy.

It is to this Act and with the organizational arrangement it legalized that our remaining efforts will be devoted. Its complete text is shown in Fig. 5.2.

Be it enacted by the Senate and House of Representatives of the United States of America in Congress assembled, That persons engaged in the production of agricultural products as farmers, planters, ranchmen, dairymen, nut or fruit growers may act together in associations, corporate or otherwise, with or without capital stock, in collectively processing, preparing for market, handling, and marketing in interstate and foreign commerce, such products of persons so engaged. Such associations and their members may make the necessary contracts and agreements to effect such purposes; Provided, however, That such associations are operated for the mutual benefit of the members thereof, as such producers, and conform to one or both of the following requirements:

First. That no member of the association is allowed more than one vote because of the amount of stock or membership capital he may own therein, or,

Second. That the association does not pay dividends on stock or membership capital in excess of 8 per centum per annum.

And in any case to the following:

Third. That the association shall not deal in the products of nonmembers to an amount greater in value than such as are handled by it for members.

SEC. 2. That if the Secretary of Agriculture shall have reason to believe that any such association monopolizes or restrains trade in interstate or foreign commerce to such an extent that the price of any agricultural product is unduly enhanced by reason thereof, he shall serve upon such association a complaint stating his charge in that respect, to which complaint shall be attached or contained therein, a notice of hearing, specifying a day and place less than thirty days after the service thereof, requiring the association to show cause why an order should not be made directing it to cease and desist from monopolization or restraint of trade. An association so complained of may at the time and place so fixed show cause why such order should not be entered. The evidence given on such a hearing shall be taken under such rules and regulations as the Secretary of Agriculture may prescribe, reduced to writing, and made a part of the record therein. If upon such hearing the Secretary of Agriculture shall be of the opinion that such association monopolizes or restrains trade in interstate or foreign commerce to such an extent that the price of any agricultural product is unduly enhanced thereby, he shall issue and cause to be served upon the association an order reciting the facts found by him, directing such association to cease and desist from monopolization or restraint of trade. On the request of such association or if such association fails or neglects for thirty days to obey such order, the Secretary of Agriculture shall file in the district court in the judicial district in which such association has its principal place of business a certified copy of the order and of all the records in the proceeding, together with a petition asking that the order be enforced, and shall give notice to the Attorney General and to said association of such filing. Such district court shall thereupon have jurisdiction to enter a decree affirming, modifying, or setting aside said order, or enter such other decree as the court may deem equitable, and may make rules as to pleadings and proceedings to be had in considering such order. The place of trial may, for cause or by consent of parties, be changed as in other cases.

The facts found by the Secretary of Agriculture and recited or set forth in said order shall be prima facie evidence of such facts, but either party may adduce additional evidence. The Department of Justice shall have charge of the enforcement of such order. After the order is so filed in such district court and while pending for review therein the court may issue a temporary writ of injunction forbidding such association from violating such order or any part thereof. The court may, upon conclusion of its hearing, enforce its decree by a permanent injunction or other appropriate remedy.

Service of such complaint and of all notices may be made upon such association by service upon any officer or agent thereof engaged in carrying on its business, or any attorney authorized to appear in such proceeding for such association, and such service shall be binding upon such association, the officers, and members thereof.

Approved February 18, 1922.

FIG. 5.2. The Capper–Volstead Act of 1922.

REFERENCES

Holman, C. W. 1982. How the Cooperative Fight Was Won. The Ionia Homestead, February 23.

Nourse, E. G. 1927. The Legal Status of Agricultural Cooperation. The Macmillan Co., New York.

Valko, L. 1981. Cooperative Laws in the U.S.A. Washington State University, Pullman, Bulletin *0902*.

TO HELP IN LEARNING

1. Prepare a short statement on the subject, "The Sherman Act was sufficient—no other legislation was needed." Relate to our objectives.

2. Discuss at a brown bag seminar with your peers this topic, "Agricultural Cooperatives have the right to be exempt from the provisions of the Sherman Act."

3. Someone has said that the public interest is served by passage of the Capper–Volstead Act. Discuss.

4. Section 2 of the Capper–Volstead Act was not needed. Discuss.

5. One of your peers has argued that special treatment of agriculture in the form of legislation is not justified. Prepare what you consider to be a logical, step-by-step response to the argument. Do this for both sides of the argument.

6. As a legislator, argue the position that legislating against monopolies may not bring the same results as enforcing free competition.

7. Prepare arguments for both sides of a debate on this subject, "Resolved that it is logical that the first step in enforcing Section 2 of the Capper–Volstead Act is to be taken by the Secretary of Agriculture."

DIRECT QUESTIONS

1. What is the significance for agriculture of the term, per se illegal?

2. What was expected of the Sherman Act?

3. Why did the Sherman Act not serve the purposes intended?

4. What is "legal collusion"?

5. What does harmonizing firm behavior with the public interest mean?

6. What is meant by countervailing power?

7. Why do you suppose agriculture was hardly mentioned in discussing the Sherman Act?

8. Why is state antitrust legislation not practical?

9. What is unique about the Federal Trade Commission?

10. What is the Federal Trade Commission Act known as?

11. Why is the Capper–Volstead Act called the Magna Carta of agriculture?

12. What does Section 1 of the Capper–Volstead Act do?

13. What does Section 2 do?

14. Where does the first step in enforcement of Section 2 take place?

15. Are agricultural cooperatives exempt from the antitrust laws?

TYING-TOGETHER QUESTIONS

1. Marketing umbrella concept, model of industrial organization, the public interest, and market power are all involved in our thinking—all have implications for where we've come in legislation. How?

2. On which of the above would you draw most heavily in preparing a case for the first step in enforcing Section 2 of the Capper–Volstead Act by the Secretary of Agriculture?

3. Are the same tools (ideas, concepts, models, etc.) used in arguing against monopoly as would be used in arguing for competition?

Part II
THE "HOW" OF AGRICULTURAL COOPERATIVES

Our efforts up to now have been aimed at developing a position in regard to the "why" of agricultural cooperatives and whether they are justified. Hopefully, this has been done or at least a great deal of progress has been made toward that end.

Let's now move to the second phase, that of concerning ourselves with the "how" of agricultural cooperatives. Let's keep in mind their why as we concern ourselves with how to make them perform in such a way that their objectives are more likely to be met.

6

Capper–Volstead Corporations and Other Types of Business— Cooperative Principles

Once the legal status of farm cooperatives had been established along with the economic bases, this form of organization was accepted as one of the four ways of doing business under the free enterprise system. Let's look a bit farther into just what type of business organization was provided for in the Capper–Volstead Act and then compare it with other types or forms used in the United States.

Before we begin this comparison, let's review some of the major aspects of the Capper–Volstead Act, especially from the standpoint of those areas relating to profits, ownership and control, and use of the cooperative as a form of business organization. It will be helpful if these distinguishing features are kept constantly in mind as we move from the why to the how of cooperatives.

Capper–Volstead makes legal an association of farmers for the purpose of marketing their products, whether incorporated or not and whether with or without capital stocks, provided:

1. It is operated for the mutual benefit of its members as producers.

2. It conforms to one or both of these requirements:
   a. No member allowed more than one vote—the often-mentioned one person–one vote principle—because of the amount of stock or membership capital owned; or
   b. Dividends on capital stock or membership capital do not exceed 8% per annum.
3. It does not deal in products of nonmembers to an amount greater in value than handled by it for its members.

It is noted that Capper–Volstead did not cover all agricultural cooperatives, only those engaged in marketing agricultural products. It is not a law under which marketing associations may be chartered. It merely permits such organizations to organize and exist without, by their very existence, being considered in violation of federal antitrust laws. It does not allow them to do things not permitted other types of organizations. They are thus subject to prosecution for unfair competition and trade practices, predatory practices of any kind, price fixing agreements with third parties, and other such activities.

Section 2 of the Capper–Volstead Act makes illegal any kind of monopolization or restraint of trade by an agricultural marketing association which results in prices of their products being unduly enhanced. Regulatory powers are given to the Secretary of Agriculture. If it is felt that a cooperative marketing association has unduly enhanced its product prices through any tactic, a cease and desist order may be issued to the cooperative. Such an order is enforceable in a federal court. If the Secretary does not act, the Justice Department can bring its own case under the Sherman, Clayton, or Federal Trade Commission Acts.

If a cooperative engages in what the court regards as predatory practices in carrying out its legitimate activities, then it is not protected from prosecution under the antitrust laws. The purchase of another business by a marketing cooperative in an above-the-board transaction would be a violation under Section 7 of the Clayton Act if the result is to substantially lessen competition or tend toward creating a monopoly. Even when a cooperative enters into lawful contracts and business activities, it may still make up a pattern of conduct unlawful under the Sherman Act.

THE ROCHDALE PRINCIPLES

The cooperative legislation we have today was influenced by the Rochdale Society, which began in England in 1844. It was a retail cooperative selling consumer goods and operated under what was referred to as certain principles. They were as follows:

1. Open membership—open to everyone.
2. One person—one vote. The person rather than the number of shares owned should be the basis for voting.
3. Cash trading.
4. Membership education.
5. Political and religious neutrality.
6. No unusual risk assumption.
7. Limited return on stock.
8. Goods sold at regular retail prices.
9. Limitation on the number of shares of stock owned.
10. Net margins (savings) distributed according to patronage.

ARE THESE BONA FIDE PRINCIPLES?
ARE THEY APPLICABLE TODAY?

If we define a principle as some sort of fundamental law, it would be correct to assume that an organization is not a bona fide cooperative if it doesn't follow the principle(s). Using this definition and applying each of the so-called principles to our cooperatives quickly reveals that most do not apply today. Since such an exercise complements our objectives in this chapter, that is, getting acquainted with the cooperative corporation as a form of business organization and comparing it with other forms, let's examine each of the principles. Should it appear that any are not fundamental to the concept of cooperatives today, they may be eliminated or renamed to something of a less crucial nature.

THE PRINCIPLES WEIGHED

Open Membership

The first so-called principle—open membership—is obviously open to question today. There is no state cooperative law requiring cooperatives to accept any and all membership applications submitted to them. A cooperative should have intelligent, informed, interested, and knowledgeably involved membership who recognize the need for a cooperative and who will contribute to its effective functioning. Applications should be received only during specified periods from producers of the relevant commodity, and they should not be automatically accepted. They should be carefully processed.

One Person–One Vote

The one person–one vote rule is designed to emphasize the service aspect of the cooperative as contrasted with the investment aspect of a business or the investor-oriented type of corporation. The basic reason for and justification of a cooperative is that it makes it possible for members to provide themselves with a service(s) which they were not able to get otherwise or the services may be provided more efficiently through this means.

Despite the importance of the service rather than the investment orientation, this rule has been modified to reflect some of the changes that have taken place in our agricultural sector. Voting plans now being used include the following:

1. One person–one vote.
2. Vote according to patronage on a dollar volume basis, but with a limitation on the number of votes that may be cast.
3. One person–one vote plus additional votes based on patronage.
4. Vote according to shares of stock (only Mississippi law permits this method).

It is recalled that the Capper–Volstead Act lists as a requirement under the Act that no member be allowed more than one vote or dividends on capital stock or membership capital should not exceed 8%.

Cash Trading

The rule regarding cash trading has little relevance today for agricultural marketing cooperatives.

Membership Education

Membership education can hardly be classified as a principle, but it is a most worthy objective. As will be discussed in a later section, understanding, interest, and knowledgeable involvement of members in their cooperative is an essential ingredient for success. The development of leadership on the part of members, especially young leadership, is also essential. The tangible and psychological benefits of communications within the cooperative and in providing an understanding of cooperatives to those outside are all educationally based.

Political and Religious Neutrality

The rule relating to political and religious neutrality takes on a special meaning in today's activist society. Farmers, by nature, have tended to leave political activity to others, but a relatively new form of action on the part of cooperatives has emerged over the past few years. Political Action Committees (PACs) have been formed by many cooperatives. Funds for them are being provided by members either on a voluntary basis or by mandatory assessment. In most cases, activity has been confined to determining the views of candidates for political office regarding agriculture in general and of cooperatives and their particular interests and problems. A basic aim has been toward educating the candidates and officeholders.

Such a movement may have been prompted by a recognition that agriculture as a percentage of the total population has grown very small. Influence could easily wane. Also, with many other groups playing strong activist roles, cooperative members may have overcome some of their past reluctance to engage in such activity. This is particularly true if they view such activities as being basically educational in nature and not completely political in the usual sense.

No Unusual Risk Assumption

No problems would usually be involved with agricultural cooperatives in avoiding unusual risks, even if this rule were considered as being a principle. Agricultural groups have tended to be rather conservative in making decisions regarding such items as expansion, facilities, and new ventures. Only in those cases where membership and boards of directors of cooperatives have not been knowledgeably involved and have to some extent abdicated their decision-making positions and roles is there evidence of unusual risk assumption.

Limited Returns on Stock

The rule relating to limited returns on stock comes very close to being a principle. This is designed to stress the service orientation of cooperative corporations and to prevent capital investment and return on investment from becoming the sole objective of membership. Many cooperatives are not organized on a stock-share basis, and interest may be paid on the money invested regardless of the form in which it is invested. The principle of limiting interest is a sound one because cooperatives are organized to serve their members as patrons—not as investors as contrasted with investor-oriented private corporations.

Many cooperatives do not pay interest on common stock, and each member usually has only one share. In the case of preferred stock where members hold varying amounts, interest is usually paid. Interest is paid on this preferred, nonvoting stock or equity because not all members have invested equally in stocks or equity.

Goods Sold at Regular Retail Prices

For several reasons, most cooperatives follow a price maintenance policy rather than selling at cost. It is difficult, if not impossible, for management to predict what costs will be well in advance, with the risk of loss of equity always possible in trying to follow an at-cost policy on pricing. There is also the possibility of retaliation by profit-type businesses, resulting in price wars if an at-cost policy is followed. In some cases with some types of cooperatives, nonmembers would be given the benefit of lower prices, and this is viewed as unfair to members. Managers prefer to pass along net savings to members at the end of the year, thus easing capital requirements.

This is not a cooperative principle, but is a business policy which may be changed from time to time as conditions warrant changes. The point is that net margins or savings belong to the members of the cooperative in any event. Whether they are passed along on a day-to-day basis in the form of at-cost pricing or in a lump sum at the end of the year as a so-called thirteenth check is a decision to be made by the board of directors.

Number of Shares of Stock Owned Is Limited

The Rochdale Principle relating to limitations on the number of shares of stock owned is viewed as a power control measure. There are nonstock cooperatives, so the rule would not apply to them. If voting is restricted by a cooperative to one person—one vote or in the variations of this practice, there is no danger of misuse of power. If interest or dividend payments are limited, no undue power influence would be possible from this source. The dangers from any of these sources appear to be illusory. Despite the fact that in some credit unions, which are not Capper–Volstead cooperatives, there are limitations on the number of shares of stock owned by members, this rule cannot be given the status of a principle.

Net Margins Distributed According to Patronage

The last rule listed as a Rochdale Principle, that net margins (savings) are distributed in accordance with patronage, is probably the most secure and everlasting of all cooperative principles. All net income of a cooperative is distributed according to the volume of business done with the cooperative. This practice is in complete accord with all the user–owner–service concepts applicable to the agricultural cooperative way of doing business. It may be that this is the only real and universal cooperative principle.

PRINCIPLES OR PRACTICES?

In relation to present-day Capper–Volstead cooperatives, the following so-called Rochdale Principles may be rejected as principles and classified as practices: (1) cash trading, (2) member education, (3) political and religious neutrality, (4) no undue risk assumption, and (5) goods sold at regular retail prices.

These may be accepted, with some degree of reservation, as being somewhere between a principle and a practice: (1) open membership, (2) one person–one vote, (3) limited interest on stock, and (4) limitation on number of shares of stock owned.

The following is a principle without reservation or qualification: net margins (savings) distributed according to patronage.

SO—WHAT IS A COOPERATIVE?

Legal definitions of a cooperative may vary depending upon the source, but most agree that a true cooperative is one that (1) provides service at cost, (2) is democratically controlled by its member–patrons, and (3) limits returns on equity capital.

With this rather detailed examination of the type of business organization made possible by the enactment of the Capper–Volstead Act, let's now compare this type with other forms used in our society. Such a comparison can be helpful in further emphasizing the service, member, user, owner, patron aspects which are so much a part of the cooperative and serve to distinguish it from other forms of business enterprise.

BUSINESS FORMS

As shown in Table 6.1, there are three forms of business organizations under the private enterprise system, with the third form—the corporation—being divided into the investor-oriented form and the

TABLE 6.1
Methods of Doing Business under Private Enterprise

Features compared	Individual	Partnership	Corporate form	
			Investor-owned corporation	Member/user-owned cooperative corporation
Who owns the business?	The individual	The partners	The stockholders	The member/users
Who uses the services?	Nonowner customers	Generally nonowner customers	Generally nonowner customers	Chiefly the member/owners
Who votes?	None necessary	The partners	Common stockholders	The member/users[a]
How is voting done?	None necessary	The partners	By shares of common stock	Usually one member—one vote or the amount of each member's business
Who determines policies?	The individual	The partners	Common stockholders and directors	The member/owners and directors
Are returns on owner-ship capital limited?	No	No	No	Yes—8% or less[a]
Who gets the operating proceeds?	The individual	The partners in proportion to interest in business	The stockholders in proportion to the number of stocks held	The member/users based on the amount of business done with the cooperative[a]

[a]Basic principles of farmer cooperation.

Source: Scroggs, C. L., Cooperative Principles and Concepts, in Proceedings, Employee and Collegiate Seminar, AIC, 1979, p. 25.

cooperative form. Features relating to use, control, and the recipients of the proceeds are compared in the table. Two of the types, the individual proprietorship and the partnership, will not be discussed here, but a discussion of the two types of corporations may serve to emphasize features that differentiate the cooperative corporation from the investor-oriented corporation.

A corporation, whether investor-oriented or cooperative, has its own legal personality created by the state government through a charter. It may acquire resources, own assets, produce and sell, incur debt, and extend credit. It may sue and be sued.

It can be formed only by strict compliance with the laws of the state in which it is being organized. A certain number of people, usually three or more, may form a corporation by filing the articles of incorporation giving required information with the Secretary of State and paying required fees.

A corporation may be organized for a specific period of time until a specific task is completed or in perpetuity. It provides for continuous existence in that death, disability, or bankruptcy of a stockholder or officer does not bring it to an end. Creditors are limited in their claims to the assets of the corporation.

INVESTOR-ORIENTED CORPORATIONS

Those who contribute to the capital of the investor-oriented type of corporation do not necessarily participate in its management or operation. Management may be concentrated in the hands of a group of experts who may own only a small portion of the outstanding stock. This is often mentioned in discussing personal commitment, or lack of it, on the part of management that has little or no ownership interests.

As shown in Table 6.1, the investor-oriented type of corporation provides goods and services to anyone who wishes to purchase them. For the most part, the customers are nonowners. It engages in activities designed to serve customers' needs, but its basic objective is to make a profit for and serve the investors.

This type of corporation is owned by those who have bought its stock. The stock was purchased with the expectation that dividends would be received and with the hope that the selling price of the shares would be greater when sold than when they were bought. This return on investment emphasis rather than the production of a needed good or service serves to differentiate the investor-oriented corporation from the cooperative corporation.

The business is owned by the stockholders, and it is they who vote at annual meetings on policies recommended by management, common stockholders, and the board of directors. The number of votes allowed for a stockholder is based upon the number of shares of stock owned.

Because the motivating force in purchasing stock of the corporation is a return on the investment, there is no legal limitation on size of the return that might be received on the capital investment. Since the stockholders own the corporation, the operating proceeds will go to them in proportion to the number of shares of stock they own.

THE COOPERATIVE CORPORATION

The cooperative corporation, as contrasted with the investor-oriented corporation, is owned by the member–patrons. It is they who use the services provided by the cooperative, and it is usually these services which they could not obtain as individuals or which were not available to them elsewhere that brought about the cooperative's formation in the first place.

The members have control over policies of the cooperative through their election of the board of directors and through voicing their positions on various issues. Voting is usually done on a one person–one vote basis, and rarely are other factors taken into account. Any deviation from the democratic principle of one person–one vote stems from differences in patronage or participation in the cooperative and not from differences in capital investment.

The service orientation of the cooperative corporation is emphasized by the limited returns permissible on ownership capital. This is usually 8% or less, and the requirement in this regard is spelled out in the first section of the Capper–Volstead Act.

Since the corporation is owned and controlled by its members, it is they who get any returns that might be realized. These go to the member–owners on the basis of their patronage or economic participation in their cooperative.

THE TWO CONTRASTED

It is obvious that the two types of corporations are different, especially in the areas of use, control, and the recipients of any net margins that might result from the business activities.

An investor-oriented business corporation is a legal entity whose owners have invested capital in the hope of making a profit. It is an organization of investors who have pooled their funds so the corporation can carry on certain activities for the purpose of providing a return on the money invested.

A farmer marketing cooperative, on the other hand, is an association of business firms, farmers, who have pooled their marketing activities in an attempt to make needed services available to themselves.

The net margins, profits, of an investor corporation are returned to its owners on the basis of investment. The net margins of a marketing cooperative are returned to its members on the basis of their use, patronage, or economic participation in their cooperative.

The generally accepted cooperative principles implied in these distinctive cooperative features are member ownership and control, nonprofit or operation at cost, and limited returns on capital. They serve to differentiate the cooperative corporation from the investor-oriented corporation. At the same time, they carry with them certain obligations and responsibilities on the part of members of a cooperative. These include management and financing. These, along with other relevant aspects of agricultural cooperatives, will be covered next as we move further into the how of cooperatives.

REFERENCES

American Institute of Cooperation. 1983. Cooperatives—What They Are. 1800 Massachusetts Ave., N.W., Washington, DC.

American Institute of Cooperation. Handbook on Cooperative Basics. 1800 Massachusetts Ave., N.W., Washington, DC.

Baarda, J. R. 1979. Cooperative Principles and the Law. Employee and Collegiate Seminar, NICE, Columbia, MO.

Knapp, J. G. 1969. The Rise of American Cooperative Enterprise— 1620–1920. Interstate Publishers and Printers, Inc., Dansville, IL.

Knapp, J. G. 1973. The Advance of American Cooperative Enterprise, 1920–1945. Interstate Publishers and Printers, Inc., Dansville, IL.

Marion, B. W., Landmark, Inc. 1972. (B) A Case Study of a Regional Farm Cooperative. Department of Agricultural Economics and Rural Sociology, The Ohio State University, Columbus.

McBride, G. 1979. Cooperatives in Our Economy. Michigan Farm Economics, No. *434*. Michigan State University, East Lansing.

Mueller, W. F. 1978. The Capper–Volstead Exemption. Working Paper as a member of the Economic Advisory Panel. National Commission for the Review of Antitrust Law and Procedure.

Scroggs, C. L. 1979. Cooperative Principles and Concepts. Employee and Collegiate Seminar, NICE, Columbia, MO.

U.S. Department of Agriculture, ACS. 1981. Agricultural Cooperatives, Pioneer to Modern. Cooperative Information Report *1*, Section 2.

U.S. Department of Agriculture, FCS. 1970. Antitrust Laws, Part III, Legal Phases of Farmer Cooperatives. FCS Information No. *70*.

U.S. Department of Agriculture, FCS. 1972. Cooperatives in Agribusiness. Educational Circular, *33*.

U.S. Department of Agriculture, FCS. 1972. Cooperatives in the American Private Enterprise System. Educational Aid No. *5*.

U.S. Department of Agriculture, FCS. 1975. Capper–Volstead Impact on Cooperative Structure. Cooperative Information Report *97*.

U.S. Department of Agriculture, FCS. 1976. Antitrust Laws, Legal Phases of Farmer Cooperatives. Information No. *100*.

U.S. Department of Agriculture, FCS. 1976. Cooperative Marketing Act—50th Anniversary, Farmer Cooperatives. Vol. *42*, No. 4.

U.S. Department of Agriculture, FCS. 1976. Legal Phases of Farmer Cooperatives. FCS Information *100*.

U.S. Department of Agriculture, FCS. 1977. Cooperatives in Agribusiness. Educational Circular *33*.

U.S. Department of Agriculture, FCS. 1978. Cooperative Facts. Cooperative Information Report No. *2*.

U.S. Department of Agriculture, FCS. 1982. Cooperative Principles and Legal Foundations. Cooperative Information Report *1*.

TO HELP IN LEARNING

1. Conduct a poll of your peers to determine what they consider to be the most significant difference between General Motors, for example, and a farm cooperative.

2. Lead a brown bag discussion with a group of cooperative leaders on the subject, "Farm Cooperatives Are Not Businesses—They Are a Way of Life."

3. Your congressperson is considering legislation which, if passed, would move cooperative corporations and other corporations closer together in handling net margins, etc. You were asked for help. Prepare an outline of what you would do to help.

DIRECT QUESTIONS

1. What do the terms "service-oriented" and "investor-oriented" mean to you?

2. What is the rationale for limiting the amount of nonmember business done by a cooperative?

3. What are predatory practices?

4. What is a principle? What is a practice?

5. Should membership in a farm cooperative be open to everyone? Why or why not?

6. Is member education a principle or is it a practice? Why your position?

7. What are PACs? What is your position regarding them? Why?

8. Why is limited returns on stocks very close to being a cooperative principle?

9. Does nonprofit operation mean zero net margin to you? Why?

10. Why is net margin distributed in accordance with patronage considered as being a principle?

11. What are Articles of Incorporation?

12. Farm cooperatives serve _____. Regular corporations serve _____.

TYING-TOGETHER QUESTIONS

1. How are the two types of corporations we are discussing so closely tied to the SCP model of industrial organization?

2. Consider—Enabling legislation is based on differences in the two forms of corporations—Enabling legislation is based upon principles involved relating to the form of corporation. Discuss what is involved in the two statements and reach a position on them.

3. How would you argue the position that the number of shares of stock owned should be voted is appropriate for all corporations?

7

Economic Feasibility
of a Cooperative

It can be said that from the standpoint of professionalism, two positions in regard to the appropriateness of agricultural cooperatives in overcoming marketing problems are completely untenable and do not fit into the clinical or diagnostic procedures suggested in previous chapters. These positions are as follows:

1. There is absolutely no place for cooperatives in solving marketing problems, or
2. Cooperatives are the answer to any and all problems in marketing.

If these two unprofessional and unscientific positions are ruled out, and yet it is apparent that marketing problems do exist, obviously the position with respect to farm cooperatives may be found somewhere between the two extremes. How to find that position in a logical manner will now be discussed.

LOOKING FOR SOLUTIONS

If the existence of a marketing problem has been recognized and at least partially defined by one or more persons who are concerned about it, it is logical to seek alternative courses of action to follow in trying to overcome the problem(s). Such alternatives would include an examination of various business organizational arrangements. This should include the cooperative corporation form as a possibility.

Once the possibility of the use of the cooperative corporation form of business in attempting to overcome the problem(s) is considered, it is necessary that the essential features of a cooperative be understood by those considering its use. What is there about this form of business which makes it different from other forms and perhaps uniquely adapted to the purpose being considered? Would a business form with these features cause it to be a better arrangement than other forms? Would such an arrangement be more likely to be successful in solving the recognized and defined marketing problems than other forms? What is a cooperative and why is it a unique form of business activity?

A cooperative is a business formed by a number of people to try to overcome some problem(s) in their marketing arrangements. Usually, it is a service or services that are not available to such people as individuals or, if available, these services are not being carried out as effectively or as efficiently as they should or could be. Such individuals wonder if they might perform these services for themselves more effectively through the cooperative process.

SIMILAR TO OTHER BUSINESSES

In several ways, cooperative corporations are similar to other forms of business. They are organized and operate in a similar manner, incorporate under the corporate laws of the state in which their main office is located, and have bylaws and Articles of Incorporation prepared in accordance with the legal requirements of the state and the federal enabling legislation, the Capper–Volstead Act. Members of the cooperative elect a board of directors. The board hires a manager and sets policy. The manager designs programs or plans of action to carry out the policy established by the board of directors.

DIFFERENCES

In three important ways, however, cooperatives are different from other forms of business. These are as follows:

1. The purpose of the cooperative is to serve its member-owners, to provide a service or services they could not otherwise have as in-

dividuals or could not have as effectively or as efficiently. It is not for the purpose of providing goods and services to others at a profit or to provide dividends or interest on invested funds.

2. The cooperative provides services at cost to its member-owners. Any net margins which are generated are distributed to the member-owners in direct proportion to their use of their cooperative—not in proportion to their investment. Dividends, if any, on capital invested in the cooperative are limited.

3. A cooperative is democratically controlled by the members. Voting is based upon membership, not on the number of shares of stock held. Usually each member has only one vote.

These basic principles or distinctive features of a cooperative should be discussed and thoroughly understood by those who are looking for possible solutions to their problems. Once this is done, such persons are in a position to make a decision as to whether they should go any further in considering a cooperative as a possible method of solving their marketing problems.

THE GROUNDWORK

If the decision is to proceed, the next step is to find out in general what is involved in starting a cooperative. This would involve finding someone who is familiar with the process of forming a cooperative to work with them in taking the necessary steps all the way through the process. Such a person can usually be found through the County Cooperative Extension Service office. Other sources of help might include state associations of farmer cooperatives and banks for cooperatives.

Once this advisory role is being performed, the leaders, with the help of their advisor, begin to compile the facts and figures necessary to present the idea of a cooperative as a possible solution to their problem to other potential members. At this stage, the process is still being handled in an informal manner, but with definite aims and objectives in mind. Professionalism, in the sense of developing factual bases for decision making and keeping emotional considerations to a minimum, should serve as a constant guiding force.

FIRST MEETING OF POTENTIAL MEMBERS

If the decision at this point is that a cooperative has potential in solving the problem(s), the next step is to determine if there is enough interest beyond the persons who had the idea in the first place to justify going any further in considering a cooperative. This can best be done at a general meeting of potential members, and the planning group should arrange such a meeting.

One of the things the original idea group and its advisor would have done is to compile a tentative list of potential members. This is the group that will be asked to attend the first meeting.

The meeting will be arranged at a date, time, and place as convenient as possible for those who will be asked to attend. Invitations to attend may be extended in all possible ways. A definite program or agenda for the meeting will be systematically planned by the group and advisor. A capable chairperson should be chosen and a business-like meeting should be held.

A very carefully prepared factual presentation of information relating to the perceived marketing problems and of the cooperative business form being proposed as a solution to the problems should be made. This should be done in a business-like manner by someone who had been involved in developing the factual information being presented. The appearance of pressure or hard sell tactics should be avoided.

This presentation should be followed by a discussion period in which everyone is encouraged to express their views and ask questions. The starting group and the advisor should be prepared to respond to all questions.

After the matter has been thoroughly discussed and all questions answered, the chairperson should determine if there is enough interest to go any further in considering the possible use of a cooperative. This can be done by asking the group for a vote as to whether the next step in the process, making a detailed study of what is involved, should be taken.

If sufficient interest to take the next step is indicated, the chairperson will appoint a committee to make a survey of all aspects of the proposed cooperative and report back to the group at a later meeting. A target date for completion of the study and the next meeting date should be agreed upon before the meeting is adjourned. Interim reports from the committee in regard to their progress should be planned.

The members of the Survey Committee should be recognized by the groups as leaders capable of making sound judgments and as having business ability. They will recognize the possible need to seek expert advice from those who have had experience with cooperatives or who have expertise in other areas. The leader group has not asked for or expected anyone to commit themselves to becoming a member of the cooperative being discussed. Many things will have been discussed, however, that the prospective members can be thinking about before the next meeting when the findings of the Survey Committee will be reported.

WORK OF THE SURVEY COMMITTEE

The work of the Survey Committee has two parts: First, it must be able to judge whether the proposed cooperative is likely to succeed in solving the recognized problems. Second, if the position of the Survey Committee is positive in regard to the first question, a specific, detailed organization pattern for the new cooperative must then be mapped out. The Committee should seek the advice of experts in fields such as law, financing, accounting, credit, economics, and engineering. It is important to recognize when such advice is needed and to obtain it. A very carefully prepared report including the recommendations of the Committee will be prepared for presentation at the next meeting. More specifically, the Committee will address these areas: (1) need for the cooperative, (2) potential membership and volume of business, (3) management skills needed and availability of potential managers with needed skills, (4) facilities needed, (5) capital needed, and (6) operating costs.

NEED FOR THE COOPERATIVE

The Committee should carefully consider each of the questions and document answers to the fullest extent possible. The need for the cooperative would be based upon the Committee's estimate of the extent to which the cooperative would provide the needed services and their costs in relation to their present cost or in relation to their costs if provided in some other manner. If it is a new service not now being provided in the area, a cost–benefit analysis should be carefully prepared.

It should be kept in mind that a cooperative is not needed unless its members will receive benefits from it which they would not otherwise receive. If it becomes clear that there is no economic need for the cooperative, the Committee should not hesitate to say so and would then go no further. To insist upon starting a cooperative when there is no economic need would do more harm than good.

MEMBERS

Once the need for a cooperative has been established, the next step is making a fairly close estimate of the number of potential members and the volume of business each member would do with the cooperative. It is essential that at least the minimum volume of business needed for efficient operation be fairly certain. Operating expenses and overhead costs must be provided for from the very beginning. In addition, a cash flow position which is satisfactory and some savings buildup are necessary.

The Committee may need to visit a sample of the potential members to determine the volume of business that would be done with the cooperative. In expanding the sample to cover all potential members and their volume of business, it is best to be conservative. Not all persons interested will join and not all who join will do so at the outset. Many prefer to wait and see how it works out before joining. Also, not all members will make fullest use of the cooperative's services.

MANAGEMENT

In most cases, a cooperative will need a full-time manager who has the necessary skills to run an efficient business. The Committee, of course, will not employ a manager, but it should make sure that if a cooperative is formed, a manager with the requisite skills would be available. They should understand and appreciate the important role of the manager in causing the cooperative to be able to perform in the way that it must if it is to meet its objectives.

FACILITIES

A very important question facing the Committee has to do with the facilities needed to make it possible for the cooperative to function properly. What land, buildings, and equipment will be needed and how much will they cost?

The type of service(s) being provided and the expected volume of business that will be done by the charter members are basic elements in making estimates as to facilities needed and their costs. Allowance for future expansion should be made. These estimates, again, should be conservative because excess capacity can prove to be very expensive.

Alternatives such as leasing an existing plant, buying used equipment, as well as building a new plant should be explored. In many cases, the advice of skilled engineers or technicians will be needed.

COSTS

Estimating operating costs of the cooperative is one of the most important jobs of the Committee. It is essential that the income–expense relationship be such that it compares favorably with the existing situation. Potential members should not be led to expect greater savings than the cooperative can realistically be expected to achieve.

All appropriate operating costs should be included in making the estimates. If per unit operating costs as estimated show little or no saving over present costs, the Committee may want to estimate the volume of business necessary to make the business worthwhile. Per unit costs tend to decrease in most businesses as volume increases.

MEMBER INVESTMENT—CAPITAL PLAN

Probably one of the first questions potential members will ask is how much money will I have to put up to get the cooperative going. The Committee must be prepared to answer this question, and to do so a capital plan must be worked out. Such a plan should include (1) whether the cooperative will be stock or nonstock; (2) an estimate of the amount of initial capital that will be needed. Possible sources and amounts or percentages from each source should be indicated; (3) a suggested plan for revolving funds for capital financing; and (4) a plan for capital reserve accumulation.

No set rule can be established for determining whether a cooperative should be organized with or without capital stock. Most states permit either form.

If the decision is to organize as a capital stock cooperative, members subscribe capital and are issued stock certificates as evidence. In most cases, one share of stock is issued to each member as evidence of membership. This is usually the voting stock. Additional shares of preferred (nonvoting) stock are issued as evidence of additional capital contribution to the cooperative.

If the cooperative is to be organized as a nonstock organization, a revolving fund certificate may be issued in the amount of each member's contribution. A membership fee is often used to raise a large part of the original capital needs.

The capital structure should be kept as simple as possible. If stock is issued, the par value should be kept low—from \$5 to \$20 is suggested. It is important that the capital needs of the cooperative be fairly accurately estimated and that its needs are met. The amount needed will depend on the volume of business the cooperative is expected to do and the type of service(s) it will provide. In any event, total capital needs will be related to the number of members and the volume of business that will be done.

INITIAL CAPITAL

Two kinds of initial capital will be needed—fixed and operating. Fixed capital is needed to purchase relatively large capital items of a fixed nature such as land, buildings, and equipment. Operating capital is needed for everyday uses to keep the business going. These include employee payrolls, paying the water and light bills, and buying supplies such as packing cartons and office supplies.

The needed initial capital can come from several sources. These include the members, the investing public who may invest capital to earn dividends, and from lending agencies such as a bank for cooperatives or a commercial bank.

Providing initial capital is a basic member responsibility. This is evidence of belief in or of good faith on the part of the members that the cooperative is needed and will be a growing concern. Without this evidence of good faith, the cooperative will be hard pressed to secure borrowed capital, since a basic question of potential lenders relates to the extent to which the members have provided initial capital. No fixed percentage of the initial capital needed that must be provided by the members can be established since circumstances can vary. An often-used thumb rule, however, is that at least 50% of the initial needs for fixed and operating capital should be provided by the members.

Each member's share of the initial capital should be based upon the use the member expects to make of the cooperative. In some cases, members may be willing and able to provide more than their share of the initial capital. This should be encouraged, but it will not entitle such members to any special privileges.

DEBT CAPITAL

In order to make clear the members' responsibility in providing initial equity capital, the estimated needs should be specified by type. Once the members' responsibilities are specified and based upon the extent to which they are met, the amount of debt capital needed will be known. As indicated, the amount of debt capital which may be obtained is directly related to how much equity capital the members are willing to provide. In most cases, evidence of members' capital contributions do not include a due date or a date by which it must be repaid. This makes it possible for the capital to be used as collateral for outside loans, so it is easy to see that the more initial capital the members supply the easier it will be to get the additional capital needed from outside sources.

The Committee should look into all possible sources of outside loans for fixed capital, taking into account the needs of the cooperatives and the policies of the lending institutions. It should then recommend the agency or agencies that can supply the capital needed under the conditions most suited to the needs of the cooperative. Sources of these loans include banks for cooperatives, commercial banks, insurance companies, and other cooperatives.

Operating loans, usually for a year or less, may be obtained from banks for cooperatives, commercial banks, and other sources. It is extremely important that loans for any purpose, long- or short-term, be suited to the needs of the cooperative. This places a special obligation on the Committee to properly assess the needs and to recommend sources of capital which most nearly meet those needs.

INTERNAL FINANCING

Once the cooperative is operating and doing business with its members, a revolving capital financing plan is most appropriate for use by the cooperative in meeting equity capital needs.

Under such a plan, a member authorizes the cooperative to use a certain amount of the money furnished to the cooperative through the member's economic participation. This may be a specified amount for each unit of product bought or sold or a specified percentage of the value of each unit. This money, sometimes called capital retains, is credited to the member's account on the cooperative's books. At the end of the year or the relevant accounting period, the member is issued a certificate in the total amount of capital retains for the period. It thus becomes the member's capital investment in the cooperative.

The capital retains go into a revolving fund which is usually used in the first years of the cooperative's existence to pay the long-term debt incurred when the cooperative started. After such debt obligations are sufficiently satisfied, the retains are returned to the members in the order in which they went into the fund.

This device serves the very basic purpose or aim of cooperatives in making it possible for the members to build up equity in their cooperative in direct proportion to the amount of business done with it. It also makes it possible for the "rule of currency" to be followed, and members who are currently using the cooperative will be supporting it. Persons who are no longer members for whatever reason can be repaid their investment. It provides the cooperative with a degree of flexibility in meeting changing conditions which bring about changed or changing financial needs.

RESERVE FUNDS

The Committee should also concern itself with a sound reserve position for the cooperative and make recommended provision for adequate reserves. This is for contingencies or unforeseen circumstances and are over and above the usual reserves established for depreciation or bad debts.

Almost inevitably poor business years may be experienced for various reasons. Reserve funds should be available for buffering such situations. New or expanded facilities may be needed from time to time. Reserves would eliminate the need for using borrowed funds and impairing capital. Reserves serve as a protection or insurance of the capital investments of members.

When reserves have built to an amount considered adequate, they can be revolved out to the members in the same way as revolving funds. There may be provisions in state cooperative laws which relate to the amount of reserves that may be held. However, this in no way suggests that adequate reserve funds are not appropriate.

OTHER ITEMS IN THE REPORT

In addition to the very basic areas previously covered relating to numbers of members, volume of business, financing, and so on, the Survey Committee should include other items in its report to the potential members at the second meeting. Among other things, it should include recommendations as to the size or number of services the cooperatives will provide, the area from which the members will come and the requirements or qualifications for membership, the membership fee, where the business will be located and what the business hours are, number of directors on the board and how they would be elected, how members will be paid for their product, how supplies and services provided by the cooperative will be priced, whether an organization agreement will be used in which potential members sign a document agreeing to belong to and patronize the proposed cooperative and provide a specified amount of initial capital, whether business will be done with nonmembers, the name of the organization, voting procedure, and other rules. Also, since there will be costs of getting organized and since some expense may have already been incurred by the Committee in its work, these costs should be estimated and the amount that each member would be assessed to cover the costs should be indicated. Such costs would include attorneys' fees, other expert advice, and filing fees.

The Committee now is ready to summarize its findings and decisions in some sort of orderly manner and prepare for its report to the prospective members at the second meeting. It should be prepared to answer each and every question that might be raised at the meeting. It might be appropriate to prepare summaries of the report for distribution at the meeting as an aid in the discussion and for later reference.

THE SECOND MEETING

After the Committee report has been finalized, the second meeting of the prospective members should be held. It should be chaired by a skilled person, and the basic agenda is the presentation of the survey report. As the report is presented, it should be discussed thoroughly, point by point. Suggestions and comments should be noted.

After the discussion, the chairperson determines if there is sufficient interest in going ahead with the cooperative. If there is not enough interest, then nothing else should be done. If enough persons want to go ahead to justify further steps, the chairperson then appoints an Organizing Committee.

ORGANIZING COMMITTEE

The Organizing Committee will probably be larger than the Survey Committee, and most of the Survey Committee members will be included on it because of the special knowledge they have gained in their work. Such committees may be established for various tasks to be performed by the Committee.

The Committee has several jobs to do. These include the following:

1. Sign up the required number of members.
2. Obtain the capital which has been subscribed and arrange for needed borrowed funds.
3. Prepare the required legal papers and file the Articles of Incorporation.
4. Arrange the first meeting of the original members.

Some form of organization agreement should be used to sign up the members. Those who are soliciting membership should have a clear and complete understanding as to how the cooperative is to work and should be business-like in their work. No promises about the cooperative should be made that cannot be fulfilled.

Once enough members have signed the organization agreement to assure the needed volume of business and the necessary capital, the capital subscriptions should be collected. A complete record should be kept and the funds should be turned over to someone who has been designated to receive them. Once the amount of borrowed funds which will be needed is known, sources of such funds should be examined further and recommendations made.

ORGANIZATION PAPERS

The Organizing Committee then prepares the legal organization papers. These must be drawn up with care in order that they meet the legal requirements of the state in which the cooperative will have its headquarters, but also to provide for the particular kind of organizations the incorporators want. Documents that will be needed include the Articles of Incorporation, bylaws, perhaps a marketing agreement, membership application, membership or stock certificate, revolving fund certificate, and meeting notices and waivers of notice.

ARTICLES OF INCORPORATION

The Articles of Incorporation state the kind of business being formed and conforms to state law. The name of the organization, the principle place of business, how long it is to last, the capital structure, the names of the incorporators, and the first officers of the association are usually shown.

BYLAWS

The bylaws provide a type of blueprint of the way the cooperative will do business. Again, there are state regulations that must be complied with, and the bylaws should also be consistent with the Articles of Incorporation. Requirements for membership, how meetings are called and conducted, how voting is done, how directors and officers are elected, their duties, their numbers, their pay, when and where the directors will meet, and the date of the fiscal year are specified in the bylaws.

The bylaws should be prepared in close cooperation with an attorney who understands cooperative law. This will assure that they conform to state laws. The Committee's role in their preparations is to make sure that they will permit the kind of actual operations the cooperative is being established to perform. Most state laws require that the bylaws be adopted by the cooperative within a certain time after the Articles of Incorporation have been filed. The Organizing Committee should arrange a meeting of the incorporators specified in the Articles for this purpose within the specified time.

MARKETING AGREEMENT

A marketing agreement should be prepared and used by all marketing cooperatives. Without such an agreement, the security of expectations necessary to make future contractual arrangements is not possible. It states the duty of the member to deliver a certain amount or percentage of product to the cooperative. It states the responsibilities of the cooperative to the members and shows methods used for meeting capital requirements of the cooperative and deductions from gross sales. In short, the duties and obligations of both parties, the cooperative and the member, are set forth. A continuing or self-renewing type of agreement is usually used. Complete details regarding the self-renewal or termination process should be spelled out.

Marketing agreements in which both parties understand what is involved are necessary to provide the cooperative with enough control over the products in which it deals to make it possible to function properly.

One of the Rochdale Principles calls for open membership. This is not at all practical for use by most agricultural cooperatives, since certain uniformity in the kind of commodity being handled and marketing methods is necessary for proper functioning of the cooperative. This makes it important that any cooperative have a properly completed membership application from every member. An application for membership, signed by the member and approved by the board of directors, constitutes legal proof of membership in a cooperative.

MEMBERSHIP CERTIFICATE

Upon acceptance into a cooperative, a member is issued a membership certificate as evidence of entitlement to all the rights, benefits, and privileges of members of the association. If revolving funds are used for capital accumulation by the cooperative, a revolving fund certificate constitutes a receipt for such capital retains as are made by the cooperative as deductions from returns from products.

ARTICLES OF INCORPORATION FILED

Finally, the Organizing Committee files the Articles of Incorporation for the cooperative, with the proper state offices making sure that all requirements are met. A fee for recording this document must be paid at the time it is filed.

As previously indicated, most states require that the bylaws of a cooperative must be adopted by a majority vote of members or stockholders. This must be done in the case of newly formed cooperatives within a specified period of time after the Articles of Incorporation are filed, usually within 30 days.

FIRST MEETING

Those persons named in the Articles of Incorporation attend the first meeting as charter members of the cooperative to adopt the bylaws. Under the law, those persons who are named are regarded as members or stockholders as soon as the Articles are filed. The Organizing Committee is responsible for arranging this meeting. A waiver of notice of first meeting is prepared and signed by those at the meeting, since no notice of meeting was sent.

A temporary chairperson conducts this first meeting at which the filing of the Articles of Incorporation is reported and a draft of the bylaws is presented. They are discussed, adopted as amended or as read, and each member signs them. The board of directors is elected at this meeting if it has not been named in the Articles of Incorporation.

BOARD OF DIRECTORS MEET

The board of directors should hold a meeting as soon as possible after the bylaws have been adopted, usually immediately after the first meeting. They take the action necessary to get the cooperative under way. These include electing the first officers of the cooperative and selecting a manager. In addition, several other actions are taken such as adopting a membership form, selecting a bank, and other seemingly routine matters, but which are essential to making the cooperative a going concern.

In addition to the selection of a competent manager, the job of acquiring a business site, buildings, machinery, and so on is very important in getting the cooperative started. The Survey Committee has done a great deal of spade work in investigating possible sites and buildings. The directors, however, will need to look into the matter very thoroughly on their own and take action.

This detailed process of determining the economic feasibility of a cooperative in overcoming perceived marketing problems may have seemed tedious and drawn out at times. However, it is a necessary process which makes it possible to take a professional position between the two extremes suggested as being completely unacceptable at the beginning of this chapter.

It should be pointed out that a variation of the process in determining the feasibility of starting a new cooperative can be used by boards of directors, managers, and member committees when mergers, expansions, joint ventures, and federations are being considered. While not guaranteeing success of such a move, this process will greatly increase the probability that whatever decision is made will be more soundly based.

Examples of a committee report, organization agreement form, marketing membership agreement, Articles of Incorporation, and other legal documents may be found in Sample Legal Documents (U.S. Department of Agriculture, 1981).

REFERENCES

Hogelund, J. A. 1982. Organizing Meat Packing Cooperatives: Recent Producer Attempts. ACS Research Report No. *11*. U.S. Department of Agriculture, ACS.

Michigan Department of Commerce. 1979. Business Corporation Act. Legislative Service Bureau, Michigan Department of Commerce.

Nord, M. 1984. Selling Cooperatives to Specific Audiences. NICE, Bozeman, MT.

Rust, I. W. Starting a Cooperative—Economic, Procedural, and Legal Considerations. Educational Circular *18*. U.S. Department of Agriculture.

U.S. Department of Agriculture, ACS. 1981. Sample Legal Documents. FCS Information No. *100*.

U.S. Department of Agriculture, ACS. 1982. Status of Bargaining Cooperatives. ACS Research Report No. *16*.

U.S. Department of Agriculture, ACS. 1983. Advising People about Cooperatives. Cooperative Information Report No. *29*.

U.S. Department of Agriculture, ACS. 1983. Farmer Cooperative Statistics, 1981. ACS Report *1*, Section 27.

U.S. Department of Agriculture, ACS. 1983. Members Make Cooperatives Go. Cooperative Information Report *12*.

U.S. Department of Agriculture, ACS. 1983. Understanding Your Cooperative. Cooperative Information Report *6*.

U.S. Department of Agriculture, FCS. 1975. The Sunkist Adventure. FCS Information *94*.

TO HELP IN LEARNING

1. Assume some marketing problem in a neighborhood with which you are familiar. Discuss it with a group of your peers along with possible ways of overcoming the problem. What are the alternatives?

2. Assume an agricultural cooperative is considered as one alternative. Prepare a short paper listing those features of a cooperative which you think are relevant to the consideration.

3. Along with two or three of your peers, visit an agricultural cooperative which is a going concern. Before your visit, prepare a list of questions you think are relevant along with areas about which you need information.

4. Prepare a short paper of a few paragraphs setting forth what service and member–owner–user relationship mean to you.

5. List what you consider, at this point, to be the five most important features of the cooperative corporation form of business enterprise.

6. A cooperative board of directors and manager are considering a merger with another cooperative. How would you advise them to go about making a decision?

DIRECT QUESTIONS

1. Why would you, if you do, consider the two polar positions, cooperatives are answers to every problem and there's no place for cooperatives anywhere, as being unprofessional?

2. What are Articles of Incorporation and bylaws?

3. What does "service at cost" and nonprofit mean to you?

4. Would it be just as well to skip the Survey Committee part of the group? Why?

5. How much of the initial capital should be provided by members in starting their cooperative? Why is this important?

6. What is preferred stock?

7. What are fixed and operating capital?

8. What is a capital retain?

9. What is a revolving fund?

10. What is meant by the "rule of currency"? Why is it important in cooperatives?

11. What are reserve funds?

12. What is provided for in a marketing agreement?

13. Why is a waiver of notice of first meeting necessary?

14. How many directors does a cooperative have, how are they elected, how long do they serve, and what are their duties?

TYING-TOGETHER QUESTIONS

1. Review your objectives in taking this course and the objectives of the course. List them.

2. Recall the structure/conduct/performance (SCP) model of industrial

organization. Using structure, prepare a short paper on the rationale for the cooperative corporation and the private corporation which seems clear to you as of now.

3. What does the concept "public interest" mean to you? Is there any place for this concept in your short paper? How?

4. In addition to being an effective means of financing a cooperative, does the use of capital retains have any special tie-in with the cooperative itself? Explain.

5. In seeking areas of thinking which you could use in developing a rationale for the existence of agricultural cooperatives, what would you use?

8

Cooperative Management Trio— Members, Directors, and Manager

In the first section of this book, we concerned ourselves with the why of agricultural cooperatives. We then contrasted this form of business organization with the other forms operating in a free enterprise economic system. In our approach to the how of agricultural cooperatives, we followed a detailed procedure designed to determine the economic feasibility of forming a cooperative in an attempt to overcome or solve perceived marketing problems. Let us assume that on the basis of our study, a cooperative corporation appeared feasible. We now have a cooperative and are faced with the task of making it perform in such a way that it is given a fair chance to overcome the problem(s) it was organized to solve.

No form of business can be expected to automatically perform in a satisfactory manner. No matter how well suited the business form is to a particular economic situation and no matter how strongly the feasibility study suggested the use of a cooperative, it is too much to expect that once formed, those who formed it can just sit back and let it work and that the marketing problem will go away.

FURTHER INTO THE HOW
OF AGRICULTURAL COOPERATIVES

Let's continue with the pursuit of the how aspect of our effort in this book. Once we have formed our cooperative, what is our role in helping it to do what it is potentially capable of doing? In short, how do we run our cooperative effectively?

We saw in Chapter 6 that the cooperative corporation differs from other business forms in several ways. In reviewing those ways, we note the users, owners, the policy establishers, and the recipients of any net operating proceeds are the members. This heavy emphasis upon the member-user-owner aspect of a cooperative puts it into a special category from the standpoint of someone's role in making it a going concern. It suggests that in at least two areas, management and financing, cooperatives are distinctly different from other business forms. This difference lies in the roles played by the members in these two areas, and it is essentially the effectiveness of the performance in these areas that determines whether the cooperative will, in fact, serve the purpose for which it was established.

This chapter will begin the study of the role of management in a cooperative as being a most strategic area. Subsequent chapters will explore the role of financing. The area of ownership is implicit in all our discussions.

THE MANAGEMENT TRIO, TEAM,
OR TRIUMVIRATE

The very legitimate use of the terms management trio, management team, or management triumvirate, perhaps applicable only to the cooperative enterprise, serves to emphasize one of the areas in which the cooperative corporation is distinctive among forms of business organizations.

Stockholders in the investor-oriented corporation own the corporation and are entitled to vote on corporate policies in accordance with the number of shares of stock they hold. They contribute to the capital, but they do not necessarily participate in its management or operation. They really do not expect to participate directly in management of the corporation because their basic purpose in buying shares of stock is to get a return on their investment in the form of dividends and/or appreciation in the value of the stocks held. Management is usually concentrated in the hands of professional managers who may have only a small portion of the outstanding shares of stocks.

MEMBER-OWNER ROLE

Cooperative corporations may resemble investor-owned corporations in outward appearance, but there are distinct differences. One of these is in the role of the member-owner in the area of management. This difference gives rise to the concept of the management triangle, depicted in Fig. 8.1.

The cooperative corporation management triangle shows the members as constituting the base or the foundation of the management team. This is significant. It reflects the strategic role which can and should be played by the owner-members of the cooperative—a role that comes about because people have formed cooperatives to obtain services which they could not get as economically, efficiently, or as effectively as individuals. They formed a legal entity to achieve an economic objective through joint participation of its members. The investment and operational risks, benefits gained or losses incurred, are shared by the members in proportion to their use of their cooperatives' services. It is democratically controlled by its members on the basis of their status as member-users and not as investors in the capital structure of the cooperative. Thus, members, as owners of the cooperative, are responsible for its management and its effectiveness in providing the services for which it was formed. Let's examine the role of members in the management triangle as related to other parts of the triangle and to the potential of the cooperative in achieving its goals.

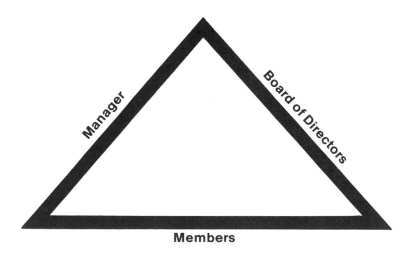

FIG. 8.1. The cooperative corporation management triangle.

Members' Management Role

Members of a cooperative have a role in management which differs from that of stockholders in the investor-oriented corporation because, as has been indicated, they own and control the cooperative. Its function is to provide needed services. In this capacity, they have certain rights and privileges in taking an active part in the management of the cooperative. This right and privilege has other facets, of course, and these include the responsibility and obligation to participate in this function in an intelligent, informed, and constructively critical manner. The success or failure of a cooperative is tied so closely to the acceptance of this responsibility and obligation and positive performance in this unique role that it deserves special attention. Membership is one of the great strengths of a cooperative if its legitimate role is effectively exercised. Positive results from this potential strength will not be realized if it is not effectively exercised.

Specific rights that come with membership in a cooperative and that reflect members' roles in management of their cooperative include the following:

1. Adopting and amending of bylaws.
2. Selecting a competent and qualified board of directors.
3. Approving plans for and changes in the capital structure of the cooperative.
4. Being knowledgeable in regard to annual reports, financial statements, and so on, and being able to ask relevant questions at all times.
5. Being familiar with the cooperative's bylaws and Articles of Incorporation, membership contracts, and other relevant documents, and requiring officers, directors, and all others to comply with them.
6. Holding directors and officers liable for any acts of omission or commission not in accord with relevant legal documents.
7. Approving all organization changes such as mergers, federations, and dissolution and all other fundamental policies of the cooperative.

It is quite obvious that if members of a cooperative are to exercise their role in managing their cooperative in an intelligent and informed manner and be able to ask meaningful questions in a constructively critical manner, it is absolutely essential that they keep themselves informed and participate in the affairs of their cooperative. They, as owners of the cooperatives, have a stake in the effectiveness with which the cooperative achieves its goals, and this is directly related to the effectiveness with which the management function is performed. Members are the beneficiaries of good management and must carry the burden of poor management.

Alert members, realizing that they, as owners of the cooperative, have definite roles to play in management, will take certain steps to assure that management performs in the best interest of the cooperative and, in doing this, performs in the best interest of the members. Certain steps may be taken which will make good management more likely. These include the following:

1. Selecting members to serve on the board of directors who are most qualified. Once elected to the board, their performance will be constantly checked.
2. Nominating and electing members to serve on the board of directors will be a procedure far removed from the "good ole boy" syndrome, but will be based on a complete understanding of the cooperative, its particular needs at the time, and whether potential nominees have the ability, knowledge, and experience to meet those needs and are willing to serve the cooperative in accordance with the nominee's qualifications and the cooperative's needs.
3. Knowing and understanding the cooperative, its reason for being, its business methods and results, and developing reasonable expectations in regard to its performance based upon knowledge and understanding.
4. Using proper channels in seeking information about the cooperative and being armed with factual information regarding it at all times.
5. Never approaching employees directly about their work or duties in carrying out administration. Any questions or suggestions should be handled through proper officials, usually the board member representing the district or area in which the member lives.
6. Being completely loyal to the cooperative and supporting it on the basis of facts and adequate information.
7. Keeping the board of directors informed in a very constructive manner on matters relating to relevant policy matters by making the views of the members known.
8. Establishing a reasonable and soundly based position in regard to length of term for directors and the number of terms which they should serve.
9. Attending meetings regularly and discussing business matters in a knowledgeable manner. Discouraging policy of allowing proxy voting.
10. Through the board of directors, allowing the manager to manage.

Allowing latitude for the manager to exercise judgment in performance of duties.

11. Accepting the position that unless members are actively and intelligently involved in their proper management role, they have no basis for complaint if the performance of the cooperative is not satisfactory.

12. Recognizing the right of the general public to ask questions about this particular cooperative and about agricultural cooperatives in general. Being prepared with factual information to explain the rationale for the existence of cooperatives, how they operate, and why they operate as they do.

Why Cooperatives Fail

One of the most frequently stated reasons for failure of cooperatives is poor management. It is generally accepted that the character and ability of management as reflected in the management function is a crucial factor in the success or failure of a cooperative.

The role of the member in the management trio is not one which is based upon direct involvement in the myriad of activities of a cooperative, such as member relations, maintaining accounting and other records, pooling, purchasing supplies, and handling of members' products. Rather, it is effectively exercised only through an adequate knowledge base regarding the cooperative and intelligent, knowledgeable participation in the affairs of the cooperative. Intelligent, constructive criticism and evaluation are necessary ingredients of this role. The legitimacy of the roles of the two sides of the management triangle stems from the member base of the triangle. Once the distinctive role of the member base is understood and adequately performed, and assuming the cooperative is soundly conceived in the beginning, the poor management reason for cooperative failure will be given far less frequently.

Let's now move to the right side of the management triangle and explore the role of the board of directors in making it possible for the cooperative to perform effectively.

THE ROLE OF THE BOARD OF DIRECTORS

The right side of the management triangle is the board of directors. This continues the total management concept which is a distinctive feature of the agricultural cooperative. It serves to emphasize further the responsibility and obligations lodged with the owners-users of an institutional arrangement that is unique as a business organization form under the free enterprise system. It reflects again the service orientation of the cooperative in which the motivation for its genesis is not a return on investment, but the provision of a service to a group of people which might not be available to them as individuals.

Selection and Election of Board Members

One of the most important roles performed by the member base of the triangle is the selection and election of the members of the board of directors. This is because the statutes under which a cooperative is established and upon which its Articles of Incorporation are based specify that the board of directors is responsible for whatever the cooperative does or does not do. This means that the board is responsible for the management of the cooperative, in accordance with the policy lines agreed upon and adopted by the member base of the management triangle.

This arrangement serves to emphasize again the importance of the process of selecting and electing members to the board because it is through the representatives on the board that members (the member base of the triangle) exercise their control over management. If this role of the member base is not effectively exercised by weighing carefully the qualifications of any member who is nominated to the board before election to serve on the board, the cooperative is likely to suffer from poor management. Nominees to the board are required to be members of the cooperative, but qualifications go far beyond membership.

Board of Directors' Responsibilities

The board of directors is empowered and obligated to run the affairs of the cooperative. They do this by (1) selecting and hiring the manager; (2) establishing policies for the cooperative; (3) delegating authority to the manager to develop programs designed to carry out policy which it establishes; (4) evaluating the effectiveness of the programs developed by the manager to carry out the policies; and (5) evaluating the performance of the cooperative and reporting to the members.

In addition to these specific responsibilities, other less tangible responsibilities of direction include recognizing and fully appreciating the fact that they are functioning as trustees for the members whom they represent. They should fully understand that they are obligated to preserve and strengthen the cooperative character of the organization, especially through encouraging member involvement and in keeping the members informed. Let's examine each of these responsibilities in more detail.

Selecting and Hiring a Manager

Selecting and hiring a competent manager is undoubtedly the most important responsibility of the cooperative's board of directors. This is because success or failure of a cooperative, as previously indicated, is so often linked to poor management.

Specifying the technical qualifications and other relevant criteria is necessary to guide the process of selecting a manager for the cooperative. Once a manager is hired, the board of directors has the responsibility to apprise the manager regarding the policies it has established for the cooperative and delegate sufficient authority to formulate programs necessary to carry out the policies. It is to be remembered that the board is responsible for everything that happens or does not happen with the cooperative. It can delegate authority to carry out programs or duties, but its responsibility cannot be delegated.

Policies and Programs

As indicated, a basic responsibility of the board of directors is to establish policies for the cooperative. Let's define what is meant by a policy.

A policy is an overall general statement of objectives of the cooperative. An example might be that the cooperative would increase its share of the market in a certain area and for a certain product. Another example would relate to the cooperative's ability to guarantee a market for its members' product over the next 10 years.

Once these policies have been established and enunciated by the board of directors, they are passed along to the manager. It is then the responsibility of the manager to devise programs to carry out the policies. In the first example, it might be by arranging for more outlets for the product in the area. In the second, it might entail the development of a long-term planning model covering estimated supplies and demand for the product over the period in question and determination of the likelihood of whether present handlers of the product for the cooperative would be adequate. Very definite steps or actions by the board may be suggested if the guarantee of market policy of the cooperative is to be honored.

Program Evaluation

In formulating the policies and articulating them to the manager for designing programs to carry them out, it may be that a reasonable period of time has been specified to operate the programs. This suggests an appraisal or evaluation of the effectiveness of the programs that the manager put into operation, and this is the responsibility of the board. Along with delegating sufficient authority to the manager to carry out the programs, the board is obligated to engage in the evaluation process as a part of the accountability posture that has to be assumed by the board.

The Manager Is Involved Also

The impression may have been given in this discussion that the manager would never be involved in the formulation of policy and that the board would follow a strictly hands-off procedure with respect to the programs implemented by the manager. This is not the case. The manager would, in most cases, be involved in an advisory capacity in the development of policy. Data and information would, in many cases, be supplied by the manager in helping the board to formulate policies.

Moreover, the board may make suggestions to the manager in regard to program alternatives in carrying out the policies. The point to be made has to do with where the responsibility lies. Policy formulation is the responsibility of the board of directors, and it cannot be delegated and should never be usurped by anyone. Programs to carry out the policies of the board are the responsibility of the manager, and sufficient authority must be delegated to carry them out. The manager should be given a free rein in doing this. The board then has the responsibility of measuring the effectiveness of the program and, depending on the evaluation, taking appropriate steps.

It is assumed in all of this, of course, that the members are performing their roles in advising the board about their views on issues. In this manner, they too are involved in policy formulation and program development and execution. The tangible part of the members' role in policy formulation is reflected in resolutions introduced and accepted at local, district, area, and annual meetings of the cooperative members or voting delegates.

Evaluation of Performance

The remaining responsibility of the board of directors is that of evaluating the overall performance of the cooperative in reaching the goals established for it and of giving an accounting of this evaluation to the members. This has implications for at least three areas. It suggests that the board of directors has a clear picture at all times of what the goals of the cooperative are and of policies designed to articulate those goals to the manager for taking action. It also implies that the members are aware of what the cooperative goals are and that they are evaluating the effectiveness of the board. Members do this in exercising their role in the management triangle. In addition, it implies that the effectiveness of the programs which have been designed by the manager to effectuate the policies or goals of the cooperative is being evaluated.

If each of these implications is, in fact, correct, then the unique roles, as depicted in the management triangle, are being exercised as they should be. The members elected a board they thought was qualified to establish the appropriate goals for the cooperative. The board accepted this mandate and took steps to specify the appropriate policies and charge the manager with designing programs to carry out the policies. The board evaluated the programs in terms of their effectiveness and, in the process, evaluated the manager. It then reported to the members in financial or other forms at the annual meeting or at other forums with respect to the cooperative's performance for a specified period. All in all, this is accountability in its best form and to a rather large measure constitutes the essence of cooperation.

Selecting Qualified Directors

If the role of the board of directors is so important in the cooperative scheme of things, their selection and election to membership on the board should be based upon their fitness for the position. Since the member base of the triangle is responsible for the makeup of the board, it is they who should sense the importance of assuring, to the fullest extent possible, that those elected to the board have the requisite qualifications. So who is qualified to serve in this critical role?

First, it is essential that the members who will be selecting persons as candidates for board positions have a complete understanding of the cooperative, its reason for being, its goals, and, of greatest importance, its particular needs at the given time. Are there particularly important issues such as expansion, merger, consolidation, integration, federation, and financing which must be faced by the board over the next few years? Rarely is there a situation or prolonged period of time when a board of directors is not called upon to make decisions for the cooperative as representatives of the member base of the management triangle. The relative importance of the issues, as measured on the basis of the consequences of the outcome of the decision, will vary, but questions of importance are constantly faced by the board of directors, and they must make decisions regarding them.

This serves to emphasize the importance of the voting members having a grasp of the workings, aims, and objectives of the cooperative and of questions that will have to be faced and resolved by the board of directors.

Second, the members, having this understanding, can then determine or make judgments regarding the particular area(s) of competence board members will need in order to address the pending issues and to enhance the chances that the decisions reached will be soundly based and in the best interest of the cooperative in the long run.

Armed with this understanding of the cooperative and its ongoing and currently important issues that must be faced and decided upon, and with an appreciation of competencies or skills needed by potential board members to properly handle the issues, the member base is in a position to make sound judgments in their director selection and election process.

Specific questions that may be asked regarding the prospective director's fitness to serve as a director might include the following:

1. Does this person have the ability to make sound business judgments as reflected by the manner in which the farm business is managed?
2. What is the track record of the person in regard to cooperatives, group action, and the like? Is there a history of working with others?
3. Has the person demonstrated leadership capacity? Is the person respected by neighbors? Is the person willing to work and put in the time that will be necessary in performing the duties of a director?
4. What particular skills does this person have which are needed by the cooperative at this time?

No one is likely to have perfect scores in this assessment procedure, but anyone who is seriously considered as a candidate for membership on the board of directors should have satisfactory scores on these and other relevant questions. To the extent that an effort is made to understand the cooperative and its needs at a particular time and to assess prospective directors' demonstrated fitness to match those needs, the chances of selecting competent persons to play the important role of cooperative director in the management triangle will be enhanced.

Directors Act as Agents of Members

In performing their role, the directors are acting as agents for the member base of the management triangle. In carrying out this agency relationship, the directors have many responsibilities to those for whom they are agents. As previously indicated, these are important, though not all inclusive:

1. Select a competent manager and delegate authority to carry out programs and perform necessary duties.

2. Evaluate effectiveness of programs and overall performance of the cooperative. Report to members.
3. Establish policies based upon relevant data and information.
4. Keep membership informed. Keep lines of communication open.

Perhaps of greatest importance is the attitudinal characteristics of the board members. The manner in which they view their work and their understanding and appreciation of what is involved in cooperative action determine in large measure the character of the cooperative. An active, capable, and enthusiastic board seems to generate enthusiasm on the part of members and goes far in assuring the proper relationship between the various parts of the management triangle. An indifferent, unenthusiastic, go-through-the-motions type of board may bring about the same posture for the members of the cooperative. It may bring about a vacuum in leadership into which a manager who is so inclined may step and create a one-person organization with a rubber stamp board. This is not cooperation, of course, and could not be expected to continue under the cooperative agenda. It will, in all likelihood, lead to failure of the cooperative.

Number of Directors and Their Terms

In determining the number of directors and their geographic location or distribution, the agency or representation role of the board of directors vis-à-vis the member base should be overriding. The board of directors represents members, and lines of communication in both directions must be kept open and logistically possible. The director–membership ratio should be such that members are assured of direct communication with a board member within a reasonable length of time. The geographic boundaries are important in determining this ratio. Accessibility of directors and open lines of communication are essential if a cooperative is to perform satisfactorily.

These considerations may also be used in determining whether some type of districting or other boundary concept is appropriate in providing for voting arrangement for members at the annual meeting. A delegate system, with delegates from a given district representing a certain number of members, may be suggested. Open, easy, accessible lines of communication and representation are hallmarks of genuine cooperation.

Committees

Depending upon the cooperative bylaws, the board is empowered to organize itself, usually after its annual meeting, and to establish committees for dealing with phases of the cooperative's business. Examples would be a finance committee, an executive committee, a marketing committee, or an auditing committee.

Each committee would be authorized by the board of directors to study problems in its particular field and make recommendations to the board of directors. Again, depending upon the bylaws, in some instances the committees may be given certain powers to act for the board, subject to review by the entire board. An executive committee made up of a specified number of members of the board performs certain duties as authorized by the board and in accordance with the legal power of the board to make such authorization.

Length of Term for Directors

In addition to taking all steps possible to assure that members who have the requisite qualifications are elected to the board, members should concern themselves with the bylaw provision which specifies the length of term a director serves. Also, the bylaws relate to whether a director may serve successive terms and how many terms may be served. The question as to whether cooperative boards of directors should rotate membership at regular intervals comes up very frequently.

Those who argue that membership on a cooperative board should be rotated periodically base their position on their feeling that this would inject new ideas, engender enthusiasms, provide greater opportunity for more members to serve the cooperative, and generally be helpful to the cooperative. They advocate a specific bylaw provision specifying that a director may not serve more than a specified number of consecutive terms and that the length of term be relatively short.

Those on the other side of the argument base their case on the too rapid removal of experience and possibly quality and that this process may be expensive to the cooperative. They argue that a period of time is needed for any member who is elected to the board to learn the workings of the cooperative and the procedures that must be followed in carrying out a director's duties and responsibilities. Automatic rotation, they argue, would result in a loss of experience and knowledge which comes only after the learning period is over, and a continuous process of learning the ropes would prevail. Unwanted or incompetent directors should be weeded out, of course, but there should be no mechanism that would automatically force a good person out of office. They argue that automatic rotation would do this, since the criterion for tenure becomes length of time rather than quality of service.

Some of the arguments of those who advocate automatic periodic rotation of directors follow:

1. New blood brings new ideas—prevents staleness.
2. Old problems may be seen in a different light by new members, and solutions may be more readily found.
3. Membership on a board is both a privilege and a burden at times, so all members should share the positive aspects of the privilege and the long hours required of conscientious board members.
4. Service as a member of the board helps to develop leadership among cooperative members, and the knowledge that a position on the board is possible will motivate young potential leaders.
5. Membership on a board of directors is one of the best teaching devices to bring about well-informed members, so more frequent openings on the board will bring about a more informed membership.
6. A definitely established policy of rotation, whether it is written into the bylaws or not, serves as a means to avoid embarrassment that may arise in unseating a director who has been in office for a long time and, for a number of reasons, may not be suited to continue.

Those who argue the other side of the question offer the following points in support of their position:

1. An experienced, able director is more valuable than an inexperienced one with equal innate ability.
2. Competency of board members is restricted by automatic rotation just as much as incompetency.
3. Automatic limitation is a restriction on the democratic right of anyone to be elected to office, including reelection of the incumbent.
4. There may not be enough qualified members to serve on the board to permit the practice of rotating directors.

We may disagree with the positions reflected by these statements on either or both sides of the argument. However, there is one position with which everyone agrees and that is that the selection of directors is one of the most important, if not the most important, aspects of the operation of a farmer cooperative.

There is much to be said for providing an opportunity for the development of potential leaders and greater understanding of cooperative philosophy and operation, but there is a reluctance to establish a system of arbitrary rotation to achieve this. Forcing good persons off the board right along with the poor ones might be a too heavy price to pay. It might be better to be left with a poor board member than to risk losing a good one. Very few cooperatives in the United States provide in their bylaws for any limitation on length of service of directors. So what should be done about this question? How can most of the recognized advantages of rotation be captured while avoiding some of the disadvantages? Cooperative leaders have offered these suggestions to achieve this goal:

1. Terms of office of directors might be staggered so that only a portion of experienced directors go off the board each year.
2. Bylaws should provide for at least two nominees for each board vacancy to be filled.
3. A nominating committee, appointed by the president of the cooperative, should present a number of nominees for the positions to be filled. No member of the board would serve on the committee. Other nominations would be accepted from the floor from the members.
4. Voting would be by secret ballot.

These suggestions may permit most of the advantages and avoid some of the disadvantages of arbitrary rotations of board members. It might work. Other plans might be devised that would work as well or better.

Associate Boards of Directors

A plan suggested by cooperative leaders which would be designed to help in the experience category for potential board members is the use of an associate board of directors.

Such a board would sit in on meetings of the board of directors, participate in the discussion of issues, and make suggestions, but, of course, would not vote.

If such a procedure were taken seriously by both the associate board members and the elected board members, potential contributions of possible future board members could be assessed. More intimate knowledge regarding the contributions being made by the legal board members could be gained by the associate members which could help them in exercising their role as responsible members in selecting directors. The length of time required by new board members to learn the ropes could be shortened. Such a plan would be very positive in the process of developing young cooperative leaders by giving potential leaders an indication of what is required to perform effectively in such a role. There are many pluses serving to recommend the use of this type of plan.

These plans, or any others, however, will work well only to the degree to which cooperative members take an interest and participate in their cooperative in an intelligent, knowledgeable manner. Given active and knowledgeable participation in their cooperative and participation in the elections, along with the procedures suggested elsewhere in this chapter regarding the needs of the cooperative and qualifications of potential members to meet those needs, a competent, representative board will almost always be selected. Lacking these necessary conditions, there is little likelihood that any plan will work.

THE COOPERATIVE MANAGER

We now move to the third part of the management triangle—the hired manager. It is an extremely important part of the triangle, and adequately filling this position is often cited as the number one responsibility of the board of directors. Its importance is reflected by the fact that with this position goes the responsibility of achieving organization objectives by effectively utilizing the resources of people, money, and materials within the policies established by the board. It is here that the programs, the means, are developed for the purpose of carrying out the objectives, the ends, or policies established by the board of directors. Its importance is further emphasized by the findings in studies of cooperatives that have failed that poor or inefficient management is the most frequently given reason for failure.

The Manager's Strategic Position

As is being stressed, the management triangle involves more than the hired manager, and weaknesses may be found in one or more of the other areas. The hired manager, however, occupies a strategic position in the triumvirate. In addition to being responsible for planning a program that will carry out the policies and objectives determined by the board of directors, the manager is responsible for hiring and directing employees to do the work. It is in the capacity of planning, organizing, directing, and controlling the affairs of the cooperative that the manager is perhaps most visible and is most often judged. It may be, however, that the more intangible capacity to reflect cooperatives' ideals and bring about a proper relationship between the three parts of the management triangle is a more fundamental test of the quality of hired management. Ability to be properly concerned with details while seeing the big picture and generally setting the cooperative's tone in a business-like setting are skills and attributes of hired management which are so important in business management. The added attitudinal attribute is a prerequisite for managing a cooperative.

Manager's Duties and Responsibilities

As has been pointed out, the hired manager is responsible for designing programs and developing the means by which the policies established by the board of directors are carried out or achieved. The details of management as a part of the management triangle rest with the hired manager and the employees. The act of combining such things as ideas, processes, materials, facilities, and people to achieve the policy goals is a direct responsibility of hired management. As this process proceeds, there is the ongoing relationship with the board of directors and the members, and it is this unique relationship in the cooperative that stamps the role of hired management as unique.

There are two ways of looking at the role of hired management in a cooperative. One is to look at what management does and the other is to look at the people involved in running the cooperative. Since the people involved in the management triangle—the members, the board of directors, and the hired manager—have already been discussed, only the functional aspects of hired management will be covered here. These relate to the usual functions of business management—planning, organizing, directing, and controlling—with only the distinctive elements that serve to differentiate cooperative business management from other types of business management being stressed.

The Functions

Planning is the thinking, judging, and deciding phase of management. It is a basic part of the process because it must precede any course of action. In most cases, there are alternative courses of action which can be taken, and an essential part of the first step in planning is the analysis of relevant data and information in deciding which alternative is most feasible. Prior to this, of course, is a thorough understanding of the policy or policies that have been formulated by the board of directors and that reflect the input of members' suggestions. The full use of the manager's experience, knowledge, and counsel would have been made by the board in formulating the policy, so the manager would not have been unaware of the policy and the underlying thinking. The close involvement of the manager and the board of directors is essential in the process of formulating policy, but once this has been done, the board should step aside and let the manager manage.

The planning function is the preparation for work and not the actual performance of the work. There is no substitute for this critical step, and it stands to reason that the better the planning, the better the following action is likely to be.

Organizing for Action

Once the planning part of the management function has been completed and the most feasible course of action has been decided upon, the organizing step takes place. This is the gearing up process of assembling the resources to carry out the plan. Personnel, facilities, and equipment will be needed and a system or systems established to perform the plan decided upon to carry out the relevant policy.

Directing Resources

Once the resources have been assembled, the next functional step of management is directing the deployment and use of the resources. This is a day-to-day execution of detailed activities in accordance with the necessary instructions. The manager is responsible for the outcome of a particular effort in relation to all other efforts. This is the process of coordination which the manager must be aware of at all times. It is here that the manager must be able to see the big picture while working with the parts.

Controlling–Evaluating

The fourth management function is that of controlling or evaluating the results being achieved in carrying out the actions designed to attain policy objectives. In this process, management determines the results which were obtained and measures them against the objectives or goals that were set up for the activity. Constant evaluation or controlling may provide bases for in-course corrections. Final assessments of the results achieved will provide bases for future activities of a similar nature, for reporting results to the board of directors and members, and for possible future policy goals.

The manager's job of supervising and coordinating various activities in the programmatic efforts to achieve policy goals covers the technical operations area and the management of human resources. Both are important and both must be done efficiently.

Duties and Responsibilities Vary

The duties and responsibilities of a manager will, of course, vary with the type of organization, the character of services rendered by it, its size, and other special conditions under which it operates. In larger cooperatives, managers' duties are largely supervisory and coordinative in character. In smaller cooperatives, the manager may be more intimately involved in carrying out the various operations. This suggests the importance of training and experience based on the requirements of the job to be done.

Selecting a Manager

As has been indicated, selecting a competent manager for a particular cooperative is one of the most important responsibilities of the board of directors.

The manager should have experience and demonstrated ability in the type of work to be done. Technical competence is essential, but it is in the area of managing human resources and in the attitudinal area of being sympathetic toward and understanding of cooperative principles and ideals that success or failure of cooperative management is really determined. The special management roles in the management triangle must be understood as one of the unique features of a cooperative and that their being performed satisfactorily is basic to cooperative success.

Member Loyalty

The extent of loyalty of the membership of a cooperative depends to a large degree upon the ability of the manager to act as a leader and to inspire confidence. The cooperative and cooperative ideals and principles must be constantly sold to all involved in cooperative activity. There is also an obligation to explain and support cooperation to the general public. These are important roles of cooperative management. They are often more difficult to perform than are the technical aspects. Cooperatives operating under entirely sound technical business principles have failed because the manager neglected or was unable to develop membership confidence and loyalty. This requirement is one that may cause difficulty when a cooperative attempts to obtain a manager from some other type of business. Some persons may find it difficult to make the necessary adjustments to manage cooperatives and may look upon the cooperative as just another business. They may not fully appreciate the implications of the fact that the owners and the patrons of the business from whom business comes are the same and that their interests are the same.

Special Relationships Must Be Kept in Mind

The implications of the previous fact and the special relationship it fosters—that reflected in the management triangle—must be fully appreciated if cooperative management is to succeed. If the board of directors is to delegate authority, the corollary is that management must accept authority. This means that all operations must be within the policies established by the board and must be designed to carry out the objectives laid down by the board. The manager must expect to be measured against the standards set by the board. Special obligations rest with the board of directors once authority to act has been delegated to the manager to let the manager manage. There is also the special understanding on the part of all the actors in the triangle that all of the authority at various levels and obligations of all parties stems from and resides in the member base of the triangle. It is clear that one of the major capacities or attributes necessary in the manager and to be reflected in the managerial role is complete understanding of and sympathy with this special cooperative feature.

REFERENCES

Bain, H. M. Legal Responsibilities of Members and Management of Cooperative Associations. U.S. Department of Agriculture, BAE, Division of Cooperative Marketing.

Bank for Cooperatives. 1977. The Cooperative Director—His Duties and Responsibilities. St. Paul, MN.

Berry, C. R., Dabney, W. T., and Voth, D. E. 1984. Managers' Perception of Member Participation in and Control of Selected Large-Scale Dairy Cooperatives. Experiment Station Bulletin *868*. University of Arkansas, Fayetteville.

Black, W. E. 1984. How Do Directors Evaluate Decisions Made and Programs Implemented? NICE, Bozeman, MT.

Dahle, R. D., and Nelson, J. L. 1973. Thinking about Cooperative Investments. North Carolina State University, Raleigh.

Duft, K. D. 1984. The Role of Directors in Strategic Long-Range Planning. NICE, Bozeman, MT.

Harris, F. L. 1984. Building and Sustaining Member Commitment. NICE, Bozeman, MT.

Hopping, R. W. 1984. Securing Management Talent. NICE, Bozeman, MT.

Knapp, J. G. A Creed for Cooperative Members. U.S. Department of Agriculture, ACS.

Lindahl, T. J., and Nelson, W. J. 1983. Working Together with the Agricultural Industry, A Cooperative Management Center, National Issues in Higher Education. Kansas State University, Manhattan.

Manuel, M. L. 1973. Improving Management of Farmer Cooperatives. General Report *120*. Farmer Cooperative Service, U.S. Department of Agriculture.

Nelson, W. J. 1984. Training and Educating Management, Directors, Employees, and Members, The Minnesota Experience. NICE, Bozeman, MT.

Rust, I. W. 1971. Should Cooperatives Rotate Directors? News for Farmer Cooperatives. Farmer Cooperative Service, United States Department of Agriculture, Washington, DC.

U.S. Department of Agriculture, ACS. 1980. Cooperative Management. ACS Report *1*, Section 8.

U.S. Department of Agriculture, ACS. 1981. Economic Impact of Two Missouri Cooperatives. ACS Research Report No. *10*.

U.S. Department of Agriculture, ACS. 1981. Measuring Cooperative Directors. Cooperative Information Report *15*.

U.S. Department of Agriculture, ACS. 1981. What Cooperative Directors Do. Cooperative Information Report *14*.

U.S. Department of Agriculture, ACS. 1982. Manager Holds Important Key to Success. Cooperative Information Report *16*.

U.S. Department of Agriculture, ACS. 1982. Member Control of Farmer Cooperatives. ACS Report No. *7*.

U.S. Department of Agriculture, ACS. 1982. Prairie Farms Dairy, Inc., Economic Impact of a Dairy Cooperative. ACS Research Report No. *12*.

U.S. Department of Agriculture, ACS. 1982. Top 100 Cooperatives, 1980 Financial Profile. ACS Research Report No. *24*.

U.S. Department of Agriculture, ACS. 1983. Cooperative Member Responsibilities and Control. ACS Report *1*, Section 7.

U.S. Department of Agriculture, ESCS. 1979. Voting Systems in Agricultural Cooperatives. Cooperative Research Report No. *12*.

U.S. Department of Agriculture, ES and CS. 1980. The Changing Financial Structure of Farmer Cooperatives. Farmer Cooperative Research Report No. *17*.

U.S. Department of Agriculture, FCS. 1973. Improving Management of Farmer Cooperatives. General Report No. *120*.

TO HELP IN LEARNING

1. Ask a group of your peers how they perceive the role of management in agricultural cooperatives.

2. Do the same with respect to private corporations. Announce your bag lunch seminar on this topic.

3. Lead a discussion in a bag lunch seminar arrangement on the

subject, "The Role of Management in Cooperative Corporations and in Private Corporations—Their Similarities and Their Differences."

4. Which part of the management triangle is, in your judgment, most important to the success of a cooperative? Why?

5. Interview a member of an agricultural cooperative and, among other things, determine feelings and views regarding the proper role of a cooperative member in management.

6. Construct a short interview questionnaire on the role of the board of directors of a cooperative. Interview in person or by telephone one or more members of a cooperative board. Analyze the responses in terms of criteria which you think are appropriate.

7. Assume you are an associate member of a board of directors of a cooperative. Attend a meeting of the board (with permission, of course). Prepare a short report for the class on your impressions and why.

8. Talk with a member of an agricultural cooperative who has never served as a member of the board. What is expected of a board member? How does the member know if expectations are being fulfilled?

9. Talk with the hired manager of an agricultural cooperative. Find out what is expected of the board. How does the manager know if expectations are being realized?

10. Discuss with an outsider, the general public, and determine the extent of knowledge of the person regarding cooperative boards of directors, what they do, how they become directors, their responsibilities, etc.

11. Discuss with a group of your peers what you feel is the most critical factor in the role of hired cooperative management.

12. Interview a cooperative manager by telephone or in person. Determine how the role of hired management in a cooperative is viewed. Determine views in regard to possible breakdowns on the part of anyone involved in the role of hired management.

13. Compare results of your interview with hired management with those obtained from interviews with members and boards of directors. What do you conclude?

DIRECT QUESTIONS

1. What is the special role of members in their cooperative?

2. What is the "management triangle"? Upon what is it based? Is there a similar situation in private corporations? Explain.

3. You are a member of an agricultural cooperative. You have a complaint about something. What do you do?

4. How long should terms of directors be?

5. Your cooperative is not functioning or performing well. To whom should you complain or ask for an explanation?

6. An outsider, the general public, asks a question about your cooperative. What is your response?

7. What role can members play in making sure that poor management doesn't cause their cooperative to fail?

8. Where can "intelligent, knowledgable, constructive criticism and evaluation" be exercised in your role as a member?

9. List a few basic general qualifications which you would consider important in selecting and electing a cooperative board member.

10. What is a policy?

11. What is a program?

12. Who is responsible for each policy and program in a cooperative? Is there ever any overlapping involvement in them by the board and manager?

13. Define evaluation. Who does it? Criteria? Why is it done?

14. Why can't a board of directors delegate responsibility? What can it delegate?

15. List a few specific qualifications which you think important in selecting a particular board of directors?

16. If you can't find someone who scores 100 on your test of qualifications, what would you then do?

17. What is likely to happen if a board of directors does not perform its duties?

18. How long should a member of a board serve? Length of term? Number of terms? Why?

19. Why are committees used by a board?

20. Would you favor having an associate board of directors? Explain your answer.

21. Define, in your own mind, what you believe management to be.

22. In this context, is management of a cooperative any different from management of a private corporation?

23. What is meant by "the big picture"? What is its relevance to the performance of a business?

24. Again, what is a policy? What is a program? Where does hired management fit?

25. How is a manager evaluated? Who does it?

26. Who does the hired manager evaluate? How is it done?

27. Someone has said that a manager should be technically competent to do everything that is done in the cooperative. Comment.

28. What is meant by the phrase, "Let the manager manage"?

TYING-TOGETHER QUESTIONS

1. Draw a "management triangle" arrangement for private corporations. Is it applicable? Explain.

2. What would you expect to happen in a situation in which cooperative members, for whatever reason(s), do not perform their management role or perform it poorly? Who takes over in such cases?

3. In view of the management role given to members, comment on the Rochdale Principle of member education which we discussed previously.

4. Describe in terms of attributes what you would consider to be an ideal cooperative member.

5. Visualize your ideal cooperative member in an annual meeting of the cooperative. What would this member do?

6. We've considered all sides of our management triangle. Which, in your judgment, is most important? Explain your answer.

7. A cooperative, for whatever reason, fails. Whose responsibility is it?

8. Discuss policy–program relationship in the context of a means–end relationship. How are goals or objectives and evaluation involved?

9. We have not mentioned our SCP model or our range of competition model for some time. Were they merely of passing interest to us? Discuss.

10. Early legislators made agricultural cooperatives legally possible. They struggled and it took time, but they did. Summarize your position in regard to what they did.

11. We've considered the third side of our cooperative management trio. What makes it unique?

12. Recall again the structural and competition elements we considered in trying to develop the why of cooperatives. Discuss the management trio concept in relationship to those considerations.

13. Discuss the job of hired management for cooperatives and for private corporations in terms of complexities, difficulties, personal satisfaction, etc.

14. Consider question (9) above and prepare a short statement on whether cooperatives are or are not a logical outgrowth and reflecption of the structural and competition elements which we considered.

9

Financing Agricultural Cooperatives

It has been indicated before that agricultural cooperatives are fundamentally different from proprietary forms of business organization in at least two areas—general management and financial management. General management, as a distinct area, was covered in the previous chapter. Let us now turn to the other area in which they are distinct—that of financial management.

The distinctiveness of these areas stems from the special nature of the cooperative regarding its reason for being, its ownership, and the recipients of any net margins which may be generated. This special nature imposes certain obligations upon the cooperative members in at least the two areas specifically mentioned.

In the case of the financial area, the obligation calls for the cooperative members to provide a substantial portion of equity capital necessary for starting the cooperative and for serving its financial requirements as it continues to operate. Equity capital is directly tied in with the potential use of debt capital in that the management of the cooperative's affairs and its capital funds is a major consideration in borrowing funds from outside sources.

Cooperatives are distinguished from other forms of business enterprise in at least these two areas, but in at least one other area they are much the same. They must be adequately financed and costs of doing business must be covered if they are to remain viable. In this sense, they are subject to the same financial requisites and required sound business decisions as are businesses of a noncooperative form.

The crucial nature of the adequacy of capital as related to the mix of equity and debt and stemming from the basic elements of cooperatives is recognized in this book by the commitment of relatively large amounts of space and time to the area. Our concern in this chapter is with a general exploration of what is involved in cooperative financing. Succeeding chapters will explore internal and external sources of financing in much greater detail. The objective is to establish firmly in the minds of the readers the necessity of adequate financing and of sound financial management. The unique characteristics of the cooperative corporation, as related to the uniqueness of agriculture itself, will be implicitly recognized at all times.

Further justification for this extensive exploration of cooperative financing lies in the fact that much of the funding, because of its nature, has special implications from a federal income tax standpoint. These will be explored in detail in a later chapter on issues relating to taxation of agricultural cooperatives.

ALL BUSINESS ENTERPRISES MUST HAVE RESOURCES

As with virtually all types of business enterprise, farmer cooperatives must have physical and human resources with which to operate. As cooperatives provide services and engage in activities which are necessary in providing those services, they must have building space, machinery, tools, equipment, trucks, automobiles, warehouses, and so on. To have control over such assets, they must have financial ability to obligate themselves for the assets and money in the bank to meet expenses.

In the process of attaining the requisite financial position for performing satisfactorily in terms of its goals and objectives, a cooperative is subject to the same fundamental principles of capital formation and accumulation as are other types of businesses. Basic to these principles is the fact that capital additions arise from savings. This means that capital formation results from withholding part of current income from current consumption and investing it in such a way that it will contribute to future production and consumption. Such withholding of savings, when properly used, will increase the capacity of the cooperative to provide the services or goods it was established to provide.

This process and the principles upon which it is grounded are very relevant to cooperative financing and should be understood by cooperative leaders and members. Savings are essential if investments in facilities and other capital items necessary in meeting the cooperative's goals and objectives are possible.

In the case of a going cooperative, the main source of capital accumulation is income diverted from immediate expenses as dividends and reinvested in permanent form in the business. As was pointed out in starting a cooperative, however, the initial capital must come from previous savings of organizers or other investors. Both the beginning and the established cooperative, however, borrow capital to supplement that which is available from reinvested savings. In this manner, credit or debt capital enters the picture. This serves to highlight the importance of maintaining a basically sound relationship between equity and debt capital and results in the use of the thumb rule relating to a 50–50 ratio between the two as being appropriate under most circumstances, especially in the case of initial capital.

A thorough understanding of what is involved in savings as the only source of investment funds, their uses, and the methods used in their accumulation can be very helpful as we move to consideration of maintaining an appropriate balance between equity and debt capital.

CAPITAL REQUIREMENTS

One of the requirements in filing the Articles of Incorporation for our cooperative was to provide evidence of paid-in capital and of a minimum payment on stock subscriptions (our cooperative is to be a stock company).

It is recalled that a rule of thumb measurement of the amount of initial capital that should be provided by the members was 50% of the amount needed to finance fixed assets and for the first year's operating requirements. Capital subscriptions were sought by the Organizing Committee in the process of assuring the needed volume of business and the capital that would be required. The subscriptions of capital were collected and turned over to someone designated by the Committee to be held for safe keeping until given to the board of directors that would be elected.

This form of capital subscription or investment in the cooperative is evidenced by shares of common or voting stock and shares of preferred or nonvoting stock. This is the method used to raise initial funds for starting a cooperative and may, of course, be used at other times to acquire equity. Capital stock may also be sold to nonmembers, but it is remembered that returns on such investments are limited and, thus, are usually at a competitive disadvantage with other investment alternatives.

It is recalled that initial capital subscribed by original members on a direct investment basis in the form of common stock served at least two purposes. First, it is a very meaningful indication of the members' faith and confidence in the eventual success of the cooperative in serving the purpose for which it was organized. It reflects their belief that the idea was soundly conceived and that the formation of the cooperative resulted from properly performed tests of economic feasibility.

The second purpose served by the willingness of the original members to provide initial capital in adequate amounts is that of providing an underpinning for debt financing which will be needed to get the cooperative under way. It is in this way that a linkage is formed between equity and debt capital. The cooperative's needs for capital cannot be met completely by debt capital. The provision of equity funds is a special obligation of members. Members, by meeting this obligation in the form of equity capital, also provide the bases for tangible and psychological collateral for use in securing debt capital.

The most likely first question to be asked by potential lenders of funds to the cooperative has to do with the amount of initial capital the original members were willing to subscribe. A logical position to take if the percentage of needed equity capital was relatively low is one of doubt as to the economic feasibility and potential viability of the cooperative. If the original members were unwilling to provide adequate capital, there is little reason to believe that others would be more willing. It is from this line of reasoning that the position that 50% of the initial equity capital should come from original members emerged as a thumb rule.

THE REMAINING CAPITAL NEEDS

If we assume that the thumb rule provision of 50% of initial capital being provided by original members by direct investment in common and preferred stock has been followed, it is quite apparent that another 50% of the capital requirements must come from elsewhere. Where will the additional capital be found?

Needed equity may be acquired by selling capital stock to nonmembers. It has also been suggested that it probably would be difficult to use this method of direct investment for needed funds because of legal limitations on rates that can be paid on earnings by a cooperative. Some funding may come from such investments, but this is not a reliable source.

Regular commercial lenders, such as banks, may also be a source of funds for the additional capital, but we recall that here, too, there are difficulties that await us.

Types of loans with respect to time involved and repayment schedules in keeping with the times when funds are available to the cooperatives to use in making payments may serve as obstacles to obtaining funds from these sources. Seasonality aspects, as they relate to storage and sales of products which impact cash flows, are all characteristics of agriculture which suggest strongly that such lending agencies have difficulty, at best, in tailoring their lending policies to meet the credit needs of farm coperatives. In some cases, such as lines of credit, these institutions may serve as credit sources, but in most cases they would have difficulty performing this role and, thus, could not be considered a major reliable source of funds. We do need additional funds and special sources have been established for this purpose.

ESPECIALLY TAILORED SOURCES
OF BORROWED FUNDS

In our coverage of the emergence of the Capper–Volstead Act, we emphasized the unique nature of agriculture as being the basic foundation of the legislation. Along with this special enabling legislation, it was also apparent that there was a need for supporting services, also based upon the uniqueness of agriculture. One of those needed supporting services related to sources of cooperative debt capital.

The need in this case was a specialized credit source which would be able to lend funds of appropriate types and with appropriate repayment schedules. Such appropriateness could come only from a thorough and sympathetic understanding of the unique credit needs of cooperatives which reflect the unique characteristics of agriculture. It is because of such an understanding that especially tailored sources of debt funds came into existence. These sources use sound, basic lending principles, but have the added dimension of understanding agriculture and tailoring their operations to fit their needs. Details of how we turn to these special credit sources will be covered later in considering external sources of financing. It is sufficient for our immediate purposes to indicate that such sources came into existence as a result of special requirements in servicing the needs of agriculture.

BACK TO OPERATING OUR COOPERATIVE

Let's assume that the operating costs of the cooperative for a year and the requirements for fixed capital had been determined by the Organizing Committee. All of this, of course, had been based upon the type of cooperative, such as a milk marketing cooperative, or a poultry cooperative, which was needed to solve the perceived marketing problems, the functions that would be performed by the cooperative, and the sales outlets and marketing methods that would be used.

Let's further assume that 50% of the needed initial capital had been subscribed on a direct investment basis by the members and that the remaining half had been secured in appropriate types of loans, appropriate advance rates, and with appropriate repayment schedules from special sources of agricultural credit, to be detailed later. The board of directors has purchased the needed land and facilities needed for the cooperative, has selected and hired a manager, and the manager has hired the needed personnel. We have equity capital from our members and debt capital from outside special sources. We must account for those funds in reporting the financial position and performance of the cooperative to the members. Let's review what is involved in some of the financial reports used for this purpose.

THE BALANCE SHEET

The board of directors has the responsibility, delegated to the manager, to prepare a statement showing the financial position of the cooperative as of an appropriate date. This statement, the balance sheet, would show the assets of the cooperative, those items owned and their values, and the liabilities of the cooperative, those items owed and their values. The difference between the two is the net worth of the cooperative or the equity of the members in their cooperative.

On the current asset side of the balance sheet, such items as cash, accounts receivable, value of inventory or stocks on hand, notes or obligations owed to the cooperative and payable within a year, and expenses that have been prepaid, such as insurance, rent, and deposits, which apply to future periods of time are listed at an appropriate value.

Assets such as buildings, land, machinery and equipment, and other investments of this type would be listed as fixed assets at a value that reflected how long they had been used, their remaining useful life, or other appropriate methods.

On the other side of the ledger, the debt items along with their values would be shown. These too would be listed as current liabilities, the obligations that are due and payable within a specified short period of time, with none running beyond a year.

These would include accounts payable, interest, social security payments, taxes, and insurance which are due shortly. Any short-term obligations to banks or others from whom funds had been borrowed and are payable within a year are included. It is important that all debts within these time categories be identified and included, since they have immediate cash flow implications.

Long-term debt or fixed liabilities would include mortgages and any other obligations payable after a year from the time the financial statement is prepared. This would be debt incurred to provide fixed assets such as land, buildings, and equipment of the cooperative. These are accounted for because they too have implications for the timely generation of funds for their payment. All debt obligations must be met, but because the unique needs of cooperatives are based upon the uniqueness of agriculture, repayment schedules may be set to recognize the nature of income flows of the borrowers.

Also shown on the liability side of the balance sheet is a type of debt or obligation which is different from those previously covered. They are nonetheless obligations of the cooperative, but because the cooperative has more latitude and discretion with respect to due dates, how discharged, and so on than with the current and long-term liabilities shown before, they are unique.

The shares of stock purchased by members in providing initial capital to get our cooperative going would be included here. This would include common stock, the usual voting stock, and preferred, nonvoting stock. Also included are membership certificates in nonstock cooperatives. All would be carried on the cooperative's books at an established par value and represent capital investment in the cooperative.

Another item included in this category is the deferred participation or patronage refunds credited on the books to members and based upon the net earnings of the cooperative and the amount of business done by each member with it. Per unit retains credited to members as a source of funds for capital purposes would be included here. Such allocations assume, of course, that the cooperative has been going for some time.

Reserves established for various purposes such as contingencies and bad debts would be accounted for. These would come from accumulation of savings, from margins on nonmember businesses, and so on. They would be retained by the cooperative, but would not be credited or allocated to the equity account of a specific member.

Any savings that had been realized and that are available for distribution or allocation to members, to reserve accounts, and so on would be a part of this special type of liability or debt obligation of the cooperative.

This type of debt or obligation covered on the balance sheet is significant from a number of standpoints, and it is important that the significance be grasped.

THE OPERATING STATEMENT

Another financial report that is essential to the management of any business or cooperative is the operating statement. The format and account entries may vary, but the purpose is the same—that of determining income and expenses and the resultant net positions—net margin, net savings, losses, and so on for a specific time period which reflect the performance of the cooperative for that period.

Revenues from the financial operation of the cooperative are based on the value of the goods or products and/or services sold. This would reflect the dollar value of sales and would be shown on the operating statement as gross sales or gross revenue. Any allowances for items such as transportation, quantity discounts, quality, cash discounts, or returned goods would be deducted from gross sales to reach a net sales position.

The goods sold, such as raw milk, were received from members and the cooperative has an obligation to pay for the goods received whether title is taken to the goods or whether the cooperative is acting only in an agency role. Whatever obligations were incurred in this category plus the value of the beginning inventory and less the value of the ending inventory will reflect the cost of goods sold. When this is subtracted from the net sales, the resultant figure is gross profit or gross margin.

Once the gross margin data have been determined, it is necessary to determine expense items which must be deducted from gross margin to find net margin or net profit. These include costs of employees' salaries and wages, costs of operating plant and equipment, costs of operating the office, and of providing office supplies. Gross margin for the period less the total of these expenses gives net margin or net profit on operations.

There may also be income from sources other than from sale of goods or from providing services. The cooperative may invest funds which it might have on hand and earn interest. Dividends may be received from some sources. Participation refunds may be received from other cooperatives from which our cooperative has borrowed funds or from which it has bought goods or services. Any expenses associated with this income would be deducted to reach a net position with respect to this nonoperating type of income. There may be other types of income in this nonoperating category, and a net position would be determined for it. A net position for all income of this type would be determined, and this too is called net profits or net savings, as was the position arrived at when income and expenses from operations were being considered previously.

Our purpose in discussing these financial instruments was to emphasize the fact that a cooperative corporation is a business enterprise much the same as other businesses: (1) It has assets and liabilities, it markets goods and/or provides services, and in the process, it has income and expenses; (2) it must have capital for operation and for longer-term fixed needs; and (3) it must have a bottom-line position over time which is positive in order to continue operations.

There is also a special significance to some of the debt or obligation items based upon the source of the item and its use in other areas, such as cooperative taxation and use as collateral for borrowed funds.

Further, there is a value to members and future members of cooperatives in being able to understand and interpret these accounting reports. The position of the cooperative as of a specific time and how it got there is shown. Usually, comparisons of positions of the cooperative at other times are shown.

The concept of the balance sheet is of interest. There must be a balance. When an entry is made on one side of the sheet, in assets or in liabilities, a similar entry must be made on the other side. In comparing a balance sheet for the year ending December 31, year 2, for example, with the year ending December 31, year 1, it is interesting to note that the income statement for the year just finished indicates how the cooperative moved from one balance sheet to the other.

Such reports reflect much more than meeting a bylaw requirement that they be prepared periodically. Interested, knowledgeably involved members will use these reports, along with all notes relating to them, as a part of the foundation for their knowledgeable involvement.

WHAT IS MEANT
BY OPERATING AT COST?

Lest we temporarily forget how cooperative corporations differ from other corporations and take the position that this is a way of doing business which is little, if any, different from other ways, let's reiterate those differences as they relate to financing. These are as follows: (1)

cooperative principles focus on operation at cost; (2) they emphasize democratic control which is generally interpreted to mean one member–one vote; and (3) they use the concept of limited returns on capital.

It has been found that these terms, referred to as basic cooperative principles, mean different things to different people. While there are inherent strengths in doing business on a cooperative basis, there are also possible weaknesses if the basic principles are not properly interpreted, especially as they relate to financing of the cooperative. The operation at cost principle or rather its interpretation is a prime example of what is involved.

Frequently, members of cooperatives interpret this as selling or pricing at cost. This, of course, has the potential of leading to cash flow problems which may prove to be disastrous. Financial realism dictates that operation at cost does not mean operating with zero cash flow as without net margins. Cooperatives, the same as any other business, must generate enough cash revenue to satisfy all of their incurred expenses and have enough cash left to service debt, revolve equity, and provide funds for future growth. The principle of financial realism must prevail, and this means that financial discipline must be practiced.

Once the principle of operation at cost is properly interpreted and financial discipline injected in accordance with this interpretation, a major strength of the cooperative corporation can be realized. This is the strength gained through linking the interest of the member–patrons, as the user of the cooperative, with the management and the capitalization of the business. The members who chose this form of business in order to get services which they could not get at all or as well otherwise and who invested their dollars in equity in the cooperative form the basis for a meaningful economic force. This potential can be realized, however, only with proper interpretation and implementation of basic cooperative principles in accordance with financial realism. With this reemphasis of unique features of the cooperative corporation, continued operation and financing of our cooperative now become our concern. Our interest will be centered on the building of an equity base in the distribution of net savings. In order to have net savings to distribute, we assume that we have interpreted the operation at cost principle in accordance with financial discipline principles and have a potentially adequate cash flow position.

CONTINUED FINANCING

It is recalled that we put our cooperative on sound financial footing when it was started by members subscribing half the initial capital needed and by borrowing the remainder from the special credit sources designed to meet the special needs of agriculture and agricultural cooperatives. All debt capital was evidenced by appropriate instruments—common and preferred stock for the members and mortgage instruments of appropriate types and repayment schedules with the bank. Our cooperative is started, financially, but now we must keep it going. Let's now examine sources of equity financing. These sources are a significant part of the consideration when a cooperative financial plan seeks to bring about an appropriate balance between its equity and debt capital.

REFERENCES

Davidson, D. R., and Street, D. W. 1983. Top 100 Cooperatives, 1982—Financial Profile. Farmer Cooperatives, United States Department of Agriculture, Agricultural Cooperative Service.

Duft, K. D. 1984. Generating Cooperative Capital at a Reasonable Cost. NICE, Bozeman, MT.

Dutrow, R. W., Brown, P. F., and Williams, R. 1981. Financial Profile of 15 New Agricultural Marketing Cooperatives. ACS Service Report No. 2.

Eerdman, H. E., and Larsen, G. H. 1965. Revolving Finance in Agricultural Cooperatives. Mimir Publishers, Inc., Madison, WI.

Engberg, R. C. 1965. Financing Farmer Cooperatives. Banks for Cooperatives, Saint Paul, MN.

Fassler, M. L. 1984. A Theoretical Justification and Application of a Current Value Accounting Model for Agricultural Cooperatives. M.S. Thesis, Department of Agricultural Economics, Michigan State University, East Lansing.

Fenwick, R. 1979. Cooperative Principles and Finance. Employee and Collegiate Seminar, NICE, Columbia, MO.

Griffin, N., Wissman, R., Monroe, W. J., Yager, F., and Perdue, E. 1980. The Changing Financial Structure of Farmer Cooperatives. FCS Research Report No. 17. U.S. Department of Agriculture.

Johnson, D. A. 1984. Cooperative Principles and Finance. NICE, Bozeman, MT.

Johnson, D. A. 1984. Financing Cooperatives. NICE, Bozeman, MT.

Nielson, K. A. 1984. Generating Capital at a Reasonable Cost. NICE, Bozeman, MT.

Rich, T. 1984. Financing Agriculture from the Long-Term Perspective. NICE, Bozeman, MT.

U.S. Department of Agriculture, ACS. 1981. Cooperative Financing and Taxation. Cooperative Information Report *1*, Section 9.

U.S. Department of Agriculture, ESCS. 1979. Financing New Cooperatives. FCS Program Aid No. 1229.

TO HELP IN LEARNING

1. Discuss the concept of a balance sheet and an operating statement with your peers as these relate to cooperative reporting. What do you conclude?

2. Prepare a brown bag seminar on the subject, "Equity and Debt Capital Mixture for Cooperatives."

3. Do the same for private corporations.

4. Summarize your findings in preparing for the two seminars.

DIRECT QUESTIONS

1. What is preferred stock?

2. What does a balance sheet show?

3. What is an operating statement?

4. Cooperative common stock is the usual _____ stock.

5. Preferred stock is _____ stock.

6. What is a contingency reserve fund? Does a cooperative need one?

7. It's a cooperative, so what's wrong with selling at cost, if anything?

8. What are equity funds?

9. What is equity capital?

10. What is debt capital?

11. A usually satisfactory ratio between the two is _____.

TYING-TOGETHER QUESTIONS

1. React to this statement, "Cooperatives are a way of life—not real business operations." Explain your reactions.

2. Are pricing or selling at cost essential to meet the nonprofit criterion for cooperatives? Explain.

3. Someone has said that an agricultural cooperative can get by with shoddy operating procedures much better than other forms of business enterprise. Comment.

4. It has been implied in all our discussions that all the unique institutional arrangements are grounded in agriculture's uniqueness. Comment.

10

Cooperative Financing— Internal Sources of Capital

We now turn to a consideration of one source of funds which may be used in generating the mixture of equity and borrowed capital deemed appropriate for the cooperative we started and are operating. Our concern here will be with equity capital as a part of that judicious mixture to be sought by every cooperative.

Emphasis upon the distinctiveness of the cooperative corporation stemming from the special threefold relationship of the member as a member, as a supplier of capital, and as an owner–patron need not be repeated, but since it is of such a basic nature with respect to agricultural cooperatives, it is recalled again. Furthermore, it is implicit in all our discussions of cooperative financing and of most other areas.

It is felt appropriate to recall this relationship because it is so apparent that the degree to which members understand and appreciate what is involved in this relationship and perform their roles in a manner that reflects this understanding largely determines whether the cooperative will perform in such a way that its objectives will be achieved. No matter how strongly a cooperative form of corporation was suggested by the feasibility study, it will not automatically do what was expected of it unless members participate knowledgeably and meaningfully in all areas, especially those relating to cooperative management and financing.

MAJOR SOURCES OF EQUITY CAPITAL

For agricultural cooperatives as a class, the major sources of equity capital as a part of the soundly conceived mixture of funds, in order of importance, are some form of retained earnings or deductions from sales made for members, the sale of equities to members of the cooperative for cash, and the sale of equities to nonmembers.

TYPE OF NET WORTH STRUCTURE NEEDED

No particular type of net worth structure has been found to be most appropriate or most applicable to all cooperatives and under all circumstances. Situations in various cooperatives will differ, and considerations regarding the net worth structure applicable to a particular cooperative will vary. In general, the considerations that are important in this decision area include (1) the amount of equity capital required in view of the nature of the operation and of other relevant factors; (2) the net worth structure, including the relative amounts of permanent and revolving capital; and (3) the method of net worth accumulation. These considerations call for competent financial planning of such a nature that the mixture of equity and borrowed capital serves present needs and has an element of flexibility built into it such that anticipated needs will be met. The equity credit capital sources that are developed, along with the related financial management policies and plans, should support a sustained ability of the cooperative to meet its objectives. Such policies and planning, of course, take into account the need for borrowing funds from external sources and the judicious mix of equity and borrowed capital which will make it possible for the cooperative to meet its present goals and objectives and those that might be anticipated.

COOPERATIVE'S NEEDS FOR CAPITAL

Cooperatives, much the same as all corporations, must have physical resources with which to operate. Manpower alone is not enough. They must have buildings and building space, machinery, tools, trucks, automobiles, warehouses, office equipment, and so on. In addition, they must have money available to meet current expenses.

It is recalled that when we were considering the feasibility of starting a cooperative, several of the major tasks assigned to the Survey Committee had to do with the capital needs of the cooperative if it is formed.

It is recalled that questions relating to facilities needed such as land, buildings, and equipment and how much they would cost were included as a part of the Survey Committee's agenda. The Committee also estimated the cost of operating the cooperative on both a total and per unit of service basis. These costs would include such items as salaries of the manager and other employees, utilities, taxes, office supplies, and other supplies needed and which would begin accumulating from the first day the cooperative started operating.

The Committee was asked to estimate the capital needs of the cooperative, both fixed or long-term, for these items and operating capital needed to keep the business going from the first day it started. It was asked to estimate how much capital would be needed to buy land, buildings, equipment, and other facilities of this type and to operate the business the first year. This was defined as initial capital, and the Committee was asked to recommend sources of the initial capital. In pursuing the question of sources of this initial capital, let's keep in mind the basic principles highlighting the fundamental differences between cooperatives and other forms of business—those that stem from the special threefold relationship—as a member, as a supplier of capital, and as a patron or user of the cooperative. These have been listed previously, and our concern here is with being constantly aware of what they are and of their significance. The interdependence of the principles and cooperative financing becomes very obvious once their significance is brought to bear in all cooperative financing considerations. This interdependence exists for all variations of cooperative structure, methods of pricing, and methods of capital accumulation.

When the methods of capital accumulation as they apply to the business community are examined, it is found that the main source of funds used to run a business is income, and the main source of capital accumulation is income diverted from immediate expenses or dividends and reinvested in the business. This, of course, applies only to going concerns. A newly organized business, such as the cooperative which we were considering, must obtain its initial capital from previous savings of organizers or other investors. The established business can, of course, borrow capital to supplement that which is available from reinvested savings. Let's return now to our cooperative and its need for capital. Keep in mind the source of funds to run the cooperative and for capital accumulation—all within the framework of the three cooperative principles as they relate to finance. Before doing this, however, let's consider a glossary of terms which are relevant in this context. These terms may be used for quick reference and should be helpful in facilitating our understanding of accumulation of equity capital. Some of the terms apply in any type of business.

GLOSSARY OF TERMS

These terms identify and define some general financial terminology and some of the terminology unique to cooperatives.

Allocated equity: The noncash refund that is credited to a member's equity account.

Allocated patronage refund: The "net" earnings of a cooperative that are returned to patrons based on the amount of business done with the cooperative during the accounting period. Both cash and noncash refunds are included.

Capital retain: A per unit assessment deducted from proceeds a member receives from pool returns. The dollar amount of the retain is added to the patron's equity in the cooperative.

Cash refund: That percentage of allocated patronage refunds distributed to the patrons in cash.

Current ratio: Current assets divided by current liabilities.

Dividend: A return paid on certain types of equity invested in the cooperative and paid independently of current patronage.

Leverage: Term debt divided by net worth on capital.

Liquidity: A measure of the ability to repay short-term debt (current liabilities on the balance sheet). One measure is the current ratio.

Noncash refund: That percentage of allocated patronage refunds retained by the cooperative as equity.

Present value: The value today of a dollar received at some point in the future. For example, the value today of $1.00 to be received 10 years from now discounted at 10% is 38.6 cents.

Solvency: A measure of the ability to repay long-term debt (long-term liabilities on the balance sheet). One measure is the net worth divided by total assets.

Unallocated equity: Earnings retained by the cooperative and not credited to the equity account of a specific member.

Revolving fund: A capital retain set up in a revolving fund category and revolved back to the member.

SOURCES OF EQUITY CAPITAL

Possible sources of the equity capital needed to start and to continue operating the cooperative include (1) the members who invest in the cooperative to get needed services, and (2) the investing public, which may invest capital in the cooperative to earn dividends.

Cooperative principles are very relevant, it is recalled, when capital needs for the cooperative are being considered along with their possible sources.

COOPERATIVE PRINCIPLES INVOLVED

The operation at cost principle has relevance for both the new business and the going concern. It is here that the cash flow concept arises, and whether the concept is interpreted as zero cash flow and without net margins or in a more realistic manner. Cooperatives, as any other business, must generate enough cash revenue to satisfy all incurred expenses and be left with sufficient funds to service debt, revolve equity, and provide funds for future growth. The policy in this regard is important to prospective members of the cooperative.

Democratic control, which is usually interpreted to mean one member–one vote, also has a bearing on the source of funds. This means that additional control through the voting procedure cannot be gained by having additional shares of stock. Therefore the service, patron, user, or participant aspect of the cooperative must be emphasized in securing the needed initial capital.

The limited returns on capital principle again serves to stress the needed service aspect of the cooperative. The basic cooperative legislation, the Capper–Volstead Act, provides for limitations on the returns on investment. This reduces or eliminates returns from stock appreciation or interest on the capital invested as a motivation to invest in the new cooperative or any other cooperative.

INITIAL CAPITAL— MEMBERS' RESPONSIBILITY

It is a basic responsibility of the members of the cooperative being formed to provide part or all of the initial capital needed. Willingness to provide needed capital is considered evidence of good faith in the cooperative enterprise being considered and of the potential members' assessment of the marketing problems being encountered and of the strength of belief that the cooperative will provide an answer to the problems.

Each member's share of the initial capital should be in proportion to the expected use of the cooperative, and both the expected use and the willingness to provide initial capital in proportion to this should have been determined by the Survey and Organizing Committees. The contribution to initial capital needs should be made in cash. Some may wish to contribute more than their proportionate share and should be allowed to do so. They should, however, be reminded of the democratic control principle of cooperatives and that their contributing more than their proportionate share will not entitle them to extra privileges in any form.

The Survey Committee and the Organizing Committee estimated total capital requirements, broke them down into fixed and operating capital, and related these to the members. This emphasizes and clarifies the member's responsibility in supplying the needed capital along with the member–supplier of capital and user relationship which has been mentioned as constituting a unique aspect of the cooperative enterprise. It also makes it easier to determine the amount of capital that will have to be borrowed and the role that debt equity will play in keeping the cooperative going.

As mentioned, the amount of capital that the cooperative can expect to borrow from outside credit sources will be very closely tied to the amount the initial members of the cooperative are willing to provide. In most cases, member capital does not have a due date or a definite date when it is to be repaid. This is referred to as permanent capital and can be used as collateral in borrowing outside funds. The more permanent capital the members supply, the easier it will be to obtain the needed outside capital.

Total capital needed by the cooperative will depend on the volume of business that will be done by the cooperative, the type of service(s) and products to be rendered, the degree and nature of competition that will be faced, and the amount of risk that will be faced in day-to-day operations. Whatever the total amount needed as worked out and estimated by the Committee, it will be related to the number of members, the type of services that will be provided, and the volume of business that will be done. In any event, a substantial portion of the initial or start-up capital should be provided by the members. To fall short of this may raise doubts regarding whether a cooperative should be started and its potential viability if it is formed.

METHODS OF
EQUITY CAPITAL ACCUMULATION

Several methods of accumulating equity capital are available for use by a cooperative. Three of these—retained patronage refunds, per unit capital retains, and base capital plans—will be discussed.

RETAINED PATRONAGE REFUNDS

A major source of equity funds for cooperatives is retained patronage refunds. As indicated in the glossary of terms, this means that a portion of the cooperative's net savings or net margin belonging to the members is retained. A portion of the total savings of the cooperative which reflects each member's use of the cooperative during a particular accounting period is allocated to the equity accounts of the members and is kept by the cooperative for use in meeting its capital requirements.

Cooperatives acquire such allocated equity from net savings left over after operating expenses and all other authorized deductions have been made from total income or revenue. This is a noncash refund that is credited to each member's equity account on the basis of patronage or participation by the member.

Retained patronage refunds, along with any cash refunds made by the cooperative, enable it to achieve the operation at cost principle. A zero net savings or zero cash flow position, even if it could be consistently achieved, would not be financially realistic, as previously discussed. Selling prices and operation costs that would achieve this result would be difficult, if not impossible, to establish ex ante. After all, members who own the cooperative also own the savings. Savings are needed to provide a flow of funds for growth and to service equity.

Noncash allocations may be issued to members as qualified or non-qualified. If they are qualified, the member to whom the allocation was made must report the amount for federal tax purposes and the cooperative deducts it from its taxable income. If the refund is nonqualified, the cooperative assumes the tax obligation and the member does not.

CASH REFUNDS

A certain percentage of allocated patronage refunds may be made to the member in cash. When cash refunds are made, the member must assume the tax liability for the refund and the cooperative is relieved of this tax liability.

Retained patronage refunds make the proportionality principle effective in that members accumulate equity in proportion to their use of the cooperative. They are subject to fluctuation with the economic ups and downs of the cooperative, since they depend upon net savings. However, they are a major source of cooperative equity.

CAPITAL RETAINS

Another major source of capital accumulation for a cooperative is the per unit capital retain plan. As shown in the glossary of terms, this is a plan in which a per unit or a percentage of sales assessment is deducted from proceeds a member receives from pool returns for products sold by the cooperative for the member. The dollar amount of the retain is added to the member's equity in the cooperative and is available for use by the cooperative in meeting its capital needs.

This source of cooperative equity is widely used by marketing cooperatives operating on a pooling basis. They are made on the basis of bylaw provisions of the cooperative or in accordance with the membership agreement signed by the member and the cooperative. They are a more stable source of cash flow for the cooperative than are retained patronage refunds, since they are not dependent on net margins. They too may be allocated as qualified or nonqualified, with the same requirements for tax purposes as for patronage refunds except that the cooperative is not required to allocate 20% of the retains in cash.

A capital retains plan has several positive features in serving the capital accumulation needs of a cooperative. It can make use of an indexing system and base the retain on a percentage of the unit value of the product sales being pooled. This takes into account changes in the per unit price of the product which would not be the case if the retain is based on a per unit volume relationship regardless of changes in prices. It too serves the proportionality concept very well.

BASE CAPITAL PLAN

Another capital formation plan which may be used by cooperatives in accumulating funds to meet their capital needs is the base capital plan. Such a plan can be tailored to fit a particular type of cooperative and its particular needs.

A base capital plan for accumulating equity capital meets the tests of currency and proportionality, which are basic cooperative principles. This comes about because such a plan for financing a cooperative is based on the position that a cooperative should determine its needs for equity capital periodically and should then adjust this base amount to fit current needs and members' use of the cooperative. Thus, a member's equity contribution to the cooperative's base capital requirements would be tied directly to the use of the cooperative, meeting in one fell swoop both the currency and proportionality standards.

Information relevant to several decision areas is required by a cooperative in considering the adoption of a base capital plan for accumulating capital to meet its needs. Most relevant, perhaps, is determining the base amount of capital the cooperative needs and an appropriate base period to be used in establishing the amount needed. Each member's share of the total base capital needed would be directly related to the member's proportionate use of the cooperative during the designated base period. Base capital needs of the cooperative would be systematically adjusted, usually on an annual basis, and each member's contribution to the total would be adjusted to fit the new base. Those members who were underinvested because their use of the cooperative was greater than was their contribution to the cooperative's base capital needs would be required to add to their contribution. Those members who were overinvested, for the reverse reason, would be credited for the overage in accordance with the systematically adjusted plan, usually receiving a cash refund of most, or all, of their excess equity. Members who have paid less than their proportionate share of the base capital requirement may be encouraged or required to make cash investments in the cooperative and pay an interest charge based on the discrepancy between their contribution to equity and what was determined to be their proportionate share of the base capital needed. They may also be encouraged to purchase equity from those members who have excess equity.

As previously indicated, the base capital plan of capital formation corresponds very closely to cooperative principles. Obligations of members to finance their cooperative are tied directly to the members' use of the cooperative. Members have the assurance in using such a plan that they are investing only their fair share, and equitable adjustments will be made if discrepancies are found between what they contributed and what their contribution should have been. The problem of handling estates of deceased and inactive members is handled forthrightly because cash payments are made in such cases in a relatively short time.

As is the usual case, however, there are also weaknesses in a base capital plan. Such plans are more complicated than most other plans, are more difficult for members to understand, and are more difficult to administer. In many cases, where cooperatives are not using a per unit retain plan, the board of directors is reluctant to increase estimated equity requirements, and the cooperative may suffer financially. In some cases too, members, especially younger persons who are more likely to be highly leveraged, may find it difficult to provide their share of the base capital needed.

ALWAYS REMEMBER

It is important to remember in considering financing that a cooperative, just as any form of business enterprise, has capital requirements that must be met if it is to succeed. Further, because of unique features of a cooperative, these needs must be met in ways that are different from other business forms. Understanding the basic role of the member–user–owner trio which is involved and the implications of this relationship for financing is essential if the cooperative's capital requirements are to be adequately met.

It is also important to remember that as in the case of most business types, responsible and successful financing of agricultural cooperatives requires a solid foundation of net worth or equity capital. This comes from the members. Earnings may be increased, growth encouraged, and better services may be provided, however, by a judicious use of borrowed funds. The best combination of borrowed funds and equity capital for a cooperative will vary according to the type of operation and other factors. There are probably success stories of cooperatives operating with a variety of financial structures based upon various combinations of equity and borrowed capital. Variations in the proportions are, of course, warranted and expected because of different sets of circumstances. We turn again, however, to our rough rule of thumb that borrowed funds should not exceed the amount of equity capital in the cooperative.

In summary, we have discussed the major plans used by cooperatives in accumulating equity capital to meet their financial requirements. Let's now move to a consideration of funds which may be used in reaching that combination of equity and borrowed capital appropriate for a cooperative based on its particular set of circumstances.

As we move to potential sources of debt capital, let's also remember that building capital and using it in accordance with sound financial planning is probably the major challenge for cooperatives today. They need capital to operate, and they need capital if they want to grow, expand, and be more able to provide needed services. That capital comes from two sources—members and debt. Most members like some form of retention of earnings to form equity capital. Debt is outside capital from nonmember investors. Cooperatives must be financially strong enough to attract debt capital and to attract it at competitive rates because the cost and availability of debt capital are based upon risk. It is an investment on the part of the lenders, made on the basis of their evaluation of the risk and the return they expect on their investment. The members' equity capital is also an investment and, to some degree, its availability is also determined by risk. There is a difference, however, in how these two classes of investors evaluate the risk and in their incentives for making the investment.

The expected return on an outside investor's investment is relatively easy to measure—it is stated simply as interest. In the case of regular commercial lenders, they compare what cooperatives are willing to pay and their perception of the risk involved against the return and risk of investing their money elsewhere.

The member–owner–user of a cooperative, on the other hand, gets a return on the investment in the cooperative in several ways. These include availability of products and services and a share of any net margins that may accumulate. It is the cooperative members' willingness to invest and patronize the cooperative, based upon their perception of the risks and returns involved, that basically determines the financial strength of the cooperative.

It is also that financial strength that the outside investors use to measure risk and determine their expectations for a return on the capital they lend to a cooperative, thus determining whether the cooperative can generate that mix of equity and debt capital which would be most appropriate for its use.

Despite the fact that special lending agencies have been developed to serve the special needs and requirements of agricultural cooperatives, cooperative boards of directors and managers must think about equity not only in terms of investment, but also in terms of risk. Lenders, even those developed to meet special requirements of agricultural cooperatives, often refer to capital, both equity and debt, as risk capital.

Let's now move to our study of sources of debt capital with greater assurance that the critical interrelationship between the two types of capital, equity and debt, will not be underestimated.

REFERENCES

Sanders, B. L. 1981. Internal versus External Funding of Cooperative Research. NICE, Bozeman, MT.
Thomas, G. A. 1980. Capitalization by Retains. NICE, Pennsylvania State University.

TO HELP IN LEARNING

1. Visit a local commercial banker and discuss the bank's lending policies with regard to cooperatives.

2. Determine if commercial banks have a rule comparable with the thumb rule that 50% of the start-up or initial capital of a cooperative should come from members.

3. Discuss in a bag lunch seminar the basic methods of accumulating equity capital for agricultural cooperatives and the rationale for each.

4. Discuss the glossary of finance terms with your peers with the view to determining their familiarity. Determine which ones have applicability only for agricultural cooperatives.

DIRECT QUESTIONS

1. A substantial portion of initial cooperative capital should come from members. Why?

2. What is an allocated patronage refund? Is this term unique to the farm cooperative?

3. Is the public usually anxious to invest in agricultural cooperatives? Why?

4. What is permanent capital?

5. What does "operation at cost" mean?

6. What do zero cash flow and no net margins mean?

7. Should goods be sold at cost? Explain.

8. What is a thirteenth check?

9. What share of the initial capital should be provided by each member?

10. Why is it so important that at least 50% of the initial capital should come from the members?

11. What is a per unit retain capital plan?

12. What is a base capital plan? What are its strengths? What are its weaknesses?

13. What is equity capital?

14. What is debt capital and how are equity and debt capital related?

TYING-TOGETHER QUESTIONS

1. About half the initial or start-up capital should be provided by the cooperative members. Trace this proposition to its economic foundations.

2. Operation at cost, democratic control, and limited returns on capital are said to be interrelated. Discuss.

3. It is said that the threefold relationship of the members of a cooperative—as a member, as a supplier of capital, and as a user of the cooperative—highlights the basic differences between cooperatives and other forms of business. Discuss.

11

Cooperative Financing—
External Sources of Capital

We have now explored sources of equity capital needed in our cooperative. It was suggested that about half of these funds—those needed for fixed capital uses such as land, buildings, and equipment, and those needed for operating the cooperative for at least the first year in paying such items as salaries, taxes, office and other supplies—should come from the members. This means that about half of the needed funds will have to come from outside or external sources. Let's now examine those sources of borrowed funds or debt capital. Although net worth generally provides the major part of their capital, cooperatives would not be able to develop their services to the fullest extent without the use of borrowed funds.

EXTERNAL SOURCES OF CAPITAL

Possible sources of borrowed funds include the investing public, which may invest capital in the cooperative to earn dividends, and loans from such lending agencies as a bank for cooperatives or a commercial bank. Each of these sources will be examined.

173

The Investing Public

It is recalled that one of the distinctive features of a cooperative related to the unique relationship stemming from the member–supplier of capital–patron concept is the limited returns on capital. This is designed to emphasize the service–user aspect of the cooperative and stresses the position that members joined together to provide for themselves a service or services they could not get at all, or as effectively, as individuals. This is in contrast, of course, with the investor-oriented corporation which aims to maximize the returns to the stockholders. It is this basic objective of service to its members that makes it distinctive and is fundamental to the very being of the cooperative. At the same time, this might be considered a disadvantage by some relative to their corporate competitors in that their ability to go to the public for funds is very limited. Investors are not interested in stocks that do not appreciate and that pay a relatively low rate of interest, as is the case with those cooperative stocks that may pay interest. Cooperatives do not have stocks to put on the market which will be highly attractive to investors.

It is possible, in some cases, especially in starting a new cooperative, for some funds to be received from other than potential members, but this cannot be depended on as a reliable source of funds. The motivation for such lending or buying of stock which has no potential for appreciation and which has a limited interest rate, practically always lower than could be received elsewhere, would have to come from other sources. In most cases, ideologically based motivation is not sufficient. We must seek other sources of such funds.

Banks for Cooperatives

It is recalled from previous chapters in Part I of this book that we developed the basic rationale for the special legislation that makes cooperative marketing possible. This legislation amended our basic antitrust legislation to permit groups of farmers to band together to market their products cooperatively without being in violation of the Sherman Antitrust Act, the Clayton Act, the Federal Trade Commission Act, or other similar legislation.

The line of reasoning used in developing the rationale for this legislation, which resulted in the Capper–Volstead Act, was structurally based. It was suggested that agriculture, although having fewer and larger farms today than a few years ago, was still the only sector of our economy which approximates the assumptions of the competitive economic model. The fact that it is still made up of a very large number of relatively small firms, as contrasted with the structure of the industries that provide inputs to agriculture and buy its output, leaves it at a power disadvantage with respect to prices and all other terms of trade. Individual farm firms were characterized as price takers and price givers whereas individual firms in industries furnishing inputs to agriculture, for example, farm machinery and those that buy agriculture's output, such as cereal processors, are price makers. This is because their industries are structured differently from that of agriculture—a small number of large firms—and each firm is thus endowed with relatively great economic power in such areas as pricing of their product, using brand names for their product, and controlling supply.

When this relatively weak economic power position of the individual farm firm is coupled with the fact that the product it produces, food, is essential for human beings, and when it is further coupled on the supply side with the risk and uncertainty of the natural elements, such as weather, insects, and diseases, it is suggested that those legislators who struggled with these very significant questions in the last half of the nineteenth century and the first part of the twentieth and came up with the special legislative acts were well advised. The special legislative acts appear to be soundly based.

The basic legislation with which we are concerned, the Capper–Volstead Act, has special provisions associated with special types of business organizations, as we have noted. These are the provisions relating to operating at cost, democratic control, and limited returns on capital. These were designed to emphasize the service aspect of the cooperative enterprise and stressed the member–supplier of capital–patron relationship that provides uniqueness and distinctiveness to the cooperative corporation as contrasted with the investor-oriented corporation.

It can be easily seen, however, that the features making the cooperative distinct and unique may, at the same time, prove to be troublesome in some of its aspects. Legislators, once aware of the troublesome areas, considered further legislation of a complementary nature which was designed to address such problem areas.

One such area was finance and credit. They reasoned that special legislation was needed in the area of credit to provide funds under such conditions and circumstances that reflect the uniqueness of agriculture as recognized in the legislation that was based upon this uniqueness. So, in addition to the Capper–Volstead Act, other legislation, including the Farm Credit Act, the Banks for Cooperatives legislation, and the Cooperative Marketing Act, was enacted. These Acts will now be examined as they relate to sources of borrowed capital for cooperatives.

THE FARM CREDIT ACT OF 1933

A major requirement for success in any business is adequate capital. This is as true with cooperative corporations as with other corporate or business arrangements. But simply appreciating the importance of adequate capital for success in agricultural cooperatives is not sufficient. Steps are needed to provide the needed capital in such a way that the unique needs of cooperative corporations are recognized based upon the distinctive features of agriculture itself and in accordance with the nature of the enabling legislation under which cooperatives operate.

COUNTRY LIFE COMMISSION

The unique credit needs of agriculture were recognized even before passage of the Capper–Volstead Act in 1922. The first explicit recognition of this uniqueness came about in 1908 when President Theodore Roosevelt established the Country Life Commission. The Commission encouraged and supported cooperatives and was instrumental in taking steps to help the poor credit position of agriculture.

For some time, recognition of the need for credit was not sufficient to bring about a coordinated effort to provide it. Only piecemeal legislative efforts were made until the Great Depression of the early 1930s drove home the need for a credit system designed to meet the peculiar needs of agriculture through their cooperatives. Early efforts to meet these needs were not overly successful and were subject to criticism on the part of many agricultural leaders. They should not, however, be considered as complete failures, because they had special merit in the necessary trailblazing efforts which finally led to the special coordinative financing systems for cooperative marketing. This is especially true of the Federal Farm Board established under the Agricultural Marketing Act approved by Congress on June 15, 1929.

FARM CREDIT ADMINISTRATION

Soon after assuming office, President Franklin Roosevelt issued Executive Order No. 6084, "Reorganizing Agricultural Credit Agencies of the United States," on March 27, 1933. The Order abolished the functions of the Federal Farm Board and changed its name to Farm Credit Administration. The name of the office of the chairman of the Federal Farm Board was changed to governor of the Farm Credit Administration. The governor was vested with all the powers and duties of the replaced Federal Farm Board.

The Executive Order of President Roosevelt was followed by a detailed law, the Farm Credit Act of 1933, and was signed into law by the president on June 16, 1933. It authorized the governor to organize and charter 12 corporations, to be known as Production Credit Corporations, and 12 banks, to be known as Banks for Cooperatives. These were to be located in each city where a Federal Land Bank had been established under previous legislation. The local Production Credit Associations were to serve the short-term credit needs of the farmers.

The credit needs of farm cooperatives were now to be served by the 12 district banks and one central bank for cooperatives. They were authorized to make loans for working capital and for facilities that were needed by the cooperatives.

The long-term credit needs of individual farmers would be directly served by the national Farm Loan Associations, their short-term credit needs would be served by the local Production Credit Associations, and the needs of their cooperatives would be served by the 12 Banks for Cooperatives or from the Central Bank for Cooperatives. The umbrella under which each of these credit agencies would operate, the Farm Credit Administration, was established as an independent government agency by the original legislation in 1933. It was transferred to the U.S. Department of Agriculture in 1939, but was moved back into an independent status in 1953. Its first governor was Henry Morgenthau, Jr., who later became Secretary of the Treasury. Its second governor was Dean Myers of Cornell University, who provided the expert knowledge of the structure of agriculture and its special credit needs for planning and constructing the newly expanded and consolidated farm credit system. A chart of the governing or control arrangement of the threefold farm credit system is shown in Fig. 11.1.

Seed capital was provided by the government for beginning the farm credit system, but Congress also established the policy that the member–borrowers should capitalize their own credit system. All capital provided by the government was returned by December 31, 1968. Thirty-five years after the organization of the Farm Credit Administration, the whole system was farmer owned.

THE FARM CREDIT ACT OF 1971

Previous credit legislation was modernized and consolidated under the Farm Credit Act of 1971. Locations of the banks in the Cooperative Farm Credit System are shown in Fig. 11.2.

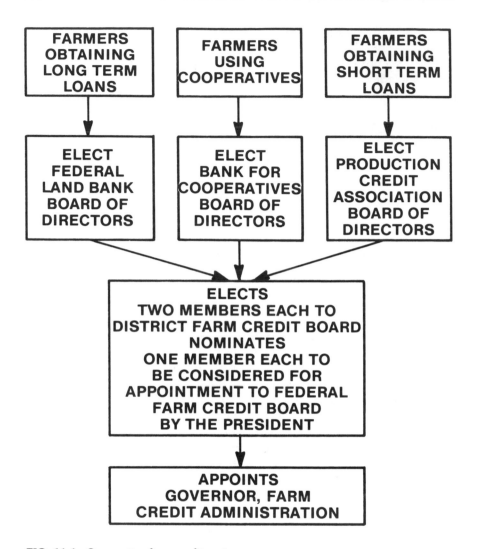

FIG. 11.1. Cooperative farm credit system.

Source: Adapted from L. Valko (1981). Cooperative Laws in the U.S.A. Bulletin 0902. Washington State University, Pullman.

The new law retained the basic principles of earlier legislation, but changes were made which were designed to broaden and modernize the service functions of the system.

Eligibility requirements for cooperatives seeking loans and services from a bank for cooperatives were changed. The requirement keeps the one person–one vote provision, but the limitation on dividends to 8% was changed by providing that an eligible cooperative does not pay

Cooperative Farm Credit System

● **FARM CREDIT ADMINISTRATION**   ■ **FARM CREDIT BANKS**
□ **FISCAL AGENCY**   **Federal Land Bank**
▲ **CENTRAL BANK FOR COOPERATIVES** **Federal Intermediate Credit Bank**
  FARMBANK SERVICE   **Bank for Cooperatives**

FIG. 11.2. Cooperative farm credit system bank locations.
Source: L. Valko (1981). Cooperative Laws in the U.S.A. Bulletin 0902. Washington State University, Pullman.

dividends on stock or membership capital in excess of what may be approved under regulations of the Farm Credit Administration. It also made it possible for rural electric or telephone cooperatives which have members who are not farmers to borrow from a Bank for Cooperatives.

The 1971 Act was amended in 1980 to allow the banks to enter the international money market through financing the agricultural exports of cooperatives.

For those wishing more details in regard to legislative activity leading up to and culminating in a credit system, especially for farm cooperatives, such information can be found in various records. Amendments relating to cooperative eligibility for borrowing and the percentage voting control held by farmers and their rationale are of interest to those who wish to become more knowledgeable in these areas. The opposition to the export financing proposal that was ultimately adopted, provided by the American Bankers Association, was vigorous and is of interest.

A basic purpose of our effort, however, relates to the first section of this book, the underlying rationale or the why of activities made possible by legislation designed to meet the particular needs of agriculture.

The legal basis of the new Farm Credit Act of 1971 and amended in 1975 is set forth in a preamble:

> It is declared to be the policy of the Congress, recognizing that a prosperous, productive agriculture is essential to a free nation and recognizing the growing need for credit in rural areas, that the farmer-owned cooperative credit system be designed to accomplish the objective of improving the income and well-being of American farmers and ranchers by furnishing sound, adequate, and constructive credit and closely related services to them, their cooperatives, and to selected farm-related business necessary for efficient farm operation.

This explicit declaration of policy by the Congress also reflects its philosophical position with respect to the public interest aspect, recognized by the legislators in providing approval for what may appear to be special treatment for agriculture. It may also be helpful to the readers in further establishing a position with respect to the why of agricultural cooperatives with which they feel comfortable and can defend.

HOW THESE BANKS WORK

As shown in Fig. 11.2, there are 12 district Banks for Cooperatives in the United States, one in each of the 12 farm credit districts. There is also a thirteenth bank, the central bank for Cooperatives, in Denver, Colorado. Each of the 12 district banks serves the credit needs of farmer cooperatives in its district. The Central Bank participates with the district banks in financing larger loans. The Banks for Cooperatives, as the name suggests, are set up to provide complete credit services to farmer cooperatives. They are owned and operated for the mutual benefit of farmer cooperatives. They are a dependable, constructive, and specialized source of credit to farm cooperatives. They are a part of the Farm Credit System which also includes the Federal Intermediate Credit Banks, the Production Credit Associations, and the Federal Land Banks, all designed to serve the unique needs of agriculture.

The Banks for Cooperatives do not lend government money, but they are closely supervised by the independent government agency, the Farm Credit Administration, discussed previously. The money loaned by the banks is obtained primarily through the sale of debentures to investors. This function is performed by the 13 banks through their agency in New York City.

The banks' only business is lending to farmer cooperatives. They understand agriculture and its unique needs for credit, are staffed by specialists who understand the unique financial characteristics and problems of agriculture, and they are a valuable source of financial counsel. They provide their borrowers, farm cooperatives, with a complete and specialized credit service adapted to their needs. Each district bank serves only the credit needs of the farmer cooperatives in its district based upon familiarity with farm business problems in its area. In addition, each bank is part of the national system. It contributes to and draws upon a national pool of information, counseling, and resources for loan funds.

Since the Banks for Cooperatives are themselves cooperatives, their borrowers share all the savings that are realized above the cost of doing business. To the extent that savings are realized, they have the effect of cutting interest costs to the farm cooperative borrower and thus to the farmer members of a cooperative.

A farm cooperative borrowing from a Bank for Cooperatives must invest in the capital of the bank. One share of stock, $100 par value, must be purchased when it obtains its first loan. It will also purchase stock in proportion to the interest paid on loans. This is Class C or voting stock. Most of the savings above the cost of doing business are distributed by the Bank for Cooperatives in the form of Class C stock. Class B, investment stock, is available to anyone who wishes to buy it.

Since the banks are cooperatives, their earnings belong to the borrowers, the users of the services of the Cooperative Banks. When the fiscal year ends, the net earnings of a bank are used to pay dividends on the Class B investment stocks. There may be other obligations and these are met. The remainder is allocated to borrowers in proportion to loan volume during the year. A portion of this may be allocated as Class C, voting stock. As the bank's financial position will permit, the oldest outstanding issue of Class C stock may be called for retirement and the owners paid in cash. Any Class B investment stock of the same year of issue, however, must first be called for retirement.

Interest rates paid by borrowers are at the lowest possible level consistent with sound operating policies. Interest is charged on the amounts advanced for the actual time the funds are outstanding to the borrower.

Interest rates are related to the cost of money in the open market. Since the money market rates change, the rates the borrowing cooperative pay can change. A clause is included by most banks in the agreement the borrower signs that interest rates may be changed. Banks usually raise or lower interest rates on outstanding loans as the cost of money in the open market increases or decreases.

TYPES OF LOANS
AND REPAYMENT SCHEDULES

A basic consideration in the rationale underlying the raison d'être for such lending institutions is the type and repayment requirements of loans which are dictated by the unique characteristics of agriculture. A frequent complaint is that regular commercial banks do not understand the characteristics of agriculture and thus are poorly equipped to make loans to farmers and their cooperatives of the type needed and with repayment schedules that are based upon these characteristics. Consequently, types of loans made to farm cooperatives by the Banks for Cooperatives and other lenders in the system and the repayment plans for the loans are tailored to fit the particular needs of farm cooperatives.

Seasonal loans or short-term capital loans are scheduled to be repaid in accordance with harvest periods, the time when funds are available. Loans to finance commodities in storage are set to be repaid from sales of the commodity being stored and held as collateral. Loans to finance facilities of a long-term nature are set up to be repaid in installments over a period of years. Payments are sometimes based on an amount for each unit of a product sold by the cooperatives. The payments may also be based on a percentage of gross sales. The overriding consideration, however, is that the type of loan and its repayment schedule reflect the unique features of agriculture and are adapted to the needs of the cooperative being financed.

SOURCES OF LOAN FUNDS

As has been indicated, a major source of funds loaned by the Banks for Cooperatives is through the sale of debentures, short-term securities in the open market to the investing public. This is one of the greatest services performed by the banks, since agricultural producers and their cooperative are not generally able to go into this market to meet their needs. Banks for Cooperatives can do this by joining together in such sales of securities. Thus, agriculture, through the Banks for Cooperatives, has access to the national money market and the advantage of low rates on large borrowings because of their sound credit ratings. The debentures, which are secured by notes of the borrowers, are joint obligations of the 13 banks. They are not guaranteed by the United States government. In carrying out these operations, the banks act as a money procurement cooperative for the farmers' cooperatives and in the process provide financing of the type needed, and with repayment schedules adapted to their needs.

Other sources of funds which are loaned by the banks to farm cooperatives include the use of their own capital which they have accumulated, regular commercial banks, and other Farm Credit Banks. They do not lend government money.

OTHER SERVICES PROVIDED

As indicated, the Banks for Cooperatives provide credit in a manner and on such terms that reflect knowledge, not only about lending principles, but of the special characteristics of agriculture and its special credit needs. In addition, they also render a general business advisory and counseling service to farm cooperatives. This is a secondary objective or service, but it is, nevertheless, important.

The banks work with farm cooperatives within their districts, whether they are stockholders or not, when requested to do so, within the limitations of their staff. They offer advice and counseling in budgeting, long-range planning, trend analyses, credit policies, and auditing standards. In some cases, the bank's attorney, when requested to do so, will work in an advisory capacity on legal matters with the cooperative's counsel.

Banks also provide consulting services to farm cooperatives when mergers and consolidations or other important steps that will affect the future of the cooperative are being considered. They are aware of the importance of the roles played by the boards of directors of cooperatives, management, membership relations, leadership development, and so on and participate in training programs designed to serve objectives relating to these areas. These services, along with the knowledgeable lending practices, give the money that is lent/borrowed greater value for both the borrower and the lender. The point is stressed that the banks are genuinely and sympathetically interested in the formation of and servicing of the credit requirements of farm cooperatives wherever the need exists and in such a manner that the cooperatives can be better served.

WHO CAN BORROW FROM BANKS
FOR COOPERATIVES?

Basic requirements for eligibility to borrow from a Bank for Cooperatives are patterned after the provisions of the Capper–Volstead Act. The cooperative seeking loan funds from a bank must be an association of farmers acting together to (1) process, prepare for market, handle, or market farm products; (2) purchase, test, grade, process, distribute, or furnish farm supplies; or (3) furnish farm business services.

The cooperative must be operated for the mutual benefit of its members and do at least 50% of its business with members. If the cooperative uses a voting plan other than one member—one vote, it cannot pay more than 8% dividends on its stock or membership capital a year. It is necessary that substantially all of the voting control of the cooperative be held by farmer members or by cooperatives owned and controlled by farmers, thus meeting the farmer-controlled criterion.

Banks stand ready to consider whether a cooperative is eligible to borrow from them. They will advise and counsel with leaders of the cooperative regarding changes which might be necessary for it to become eligible.

STEPS IN APPLYING FOR A LOAN

In applying for a loan from a Bank for Cooperatives, these steps are usually followed:

1. Officials of the cooperative contact the bank and list or outline its needs for credit with an application for such credit.
2. The bank official examines the application. If the cooperative has met the eligibility requirements and the application is approved, a loan agreement and other necessary legal papers are sent to the cooperative.
3. A representative of the bank visits the cooperative to analyze its financial situation and to inspect the facilities which would serve as collateral for the loan.
4. Authorized officials of the cooperative sign the loan documents on behalf of the cooperative and return them to the bank. Loan funds are then made available to the cooperative in accordance with the terms agreed to by the cooperative and the bank.

It has been indicated before that the Banks for Cooperatives are lending agencies much the same as regular corporate banks. They are federally chartered. They have credit standards which they use in making loans to farm cooperatives that otherwise meet the eligibility requirements for borrowing from one of the banks.

Before a loan is made to a cooperative, the bank will analyze the economic need for the loan and determine the cooperative's ability to repay the loan. The bank will satisfy itself that the cooperative has an organization, management, financial condition, and business policies as will reasonably assure its success and then assure repayment of the loan. The economic need for the cooperative's services, its membership support, its capital structure, and operating ability are all taken into account in deciding whether the likelihood that the cooperative will be able to repay the loan is great enough to warrant the bank's extension of the loan funds.

Once the first loan obligation has been successfully handled by the cooperative and its credit worthiness has been established, repeat loans are handled very expeditiously. The cooperative usually needs only to request, fill out, and submit forms for the bank's consideration and prompt action is usually taken to provide funds needed in a timely manner.

It is well to remember that the Banks for Cooperatives are cooperatives themselves and lending to farmer cooperatives is their only business. Their borrowers share all the savings that are realized over the cost of doing business. They do not lend government money—they go into the open market and sell debentures, short-term securities, to the investing public. This constitutes a very valuable service to cooperatives and to agriculture, since they are not generally able to go into this market to meet their credit needs.

FARM CREDIT SYSTEM
FACES HARDSHIPS

Perhaps no institutional arrangement can expect to have infinitely smooth sailing. This is true of the Farm Credit System despite the fact that it came into existence to meet the unique credit requirements of cooperatives and through them, the requirements of agriculture. This uniqueness provided the rationale for the Capper–Volstead Act and exposed the situation and the felt need from which the Banks for Cooperatives and the Farm Credit System sprang. In a very real sense, the old adage that "necessity is the mother of invention" applied in this case.

Despite the soundness of its foundation, the Farm Credit System came upon hard times because agriculture came upon hard times. The situation reached an economic crisis point in 1985 and late in the year Congress passed and the president signed legislation designed to reorganize and rescue the Farm Credit System, which had served agriculture's needs for many years.

WHAT HAPPENED?

It was easy to carry out the mission of the Farm Credit System throughout the 1970s. Favorable interest rates and easy terms, foreign markets armed with American dollars, incentives to increase production, and tax benefits to invest in farm equipment all combined to place farmers and farming in a very favorable profit position.

Fierce competition for land became a way of life. Farmland was doubling in value every 5 years or so. Caught up in the maelstrom, most groups including farmers came to believe that the only way to stay even was to leverage as high as possible—to buy everything possible on credit and pay for it in inflated, cheaper dollars.

But this situation didn't last. The 1980s brought a different situation. Interest rates rose and deflation set in. Our strong dollar position priced us out of the European markets. Farm prices dropped significantly. Land values dropped—50 to 75% in many cases. Balance sheet assets used as collateral for borrowing were drastically eroded.

A basic feature of this farm financial slump, the worst since the Great Depression, is the inability of farmers to service their debt which was built up on the booming land prices. In the decade ending in 1981, the average price of farmland rose from $300 per acre to over $1700. Farm debt in the United States rose from $50 billion to $200 billion. Cash farm income dropped to around $20 billion. A debt burden existed on which interest payments alone ran at $20 billion a year. The slump in land prices left many farmers who borrowed when land was dearest with farm values far below the amount of their debts. A liquidity crisis of critical proportions had developed.

MEANING FOR
THE FARM CREDIT SYSTEM

We have been discussing the economic woes of the U.S. farm sector, but what does this have to do with the Farm Credit System?

Farm sector economic problems are translated directly to the Farm Credit System because the System accounts for over 40% of all U.S. farm mortgage loan volume. When farmers face harsh financial stress resulting in asset value deterioration and liquidation, problems in getting credit, and even bankruptcy, the Farm Credit System also experiences financial stress. The System reported a loss of $522.5 million in its quarterly report for the July–September 1985 quarter and indicated that it foresaw problems at least through 1987. Losses of $2.5 billion were projected for the fourth quarter of 1985 and of about $1.5 billion for each of the next 2 years. An estimate of $10 billion or more of nonearning assets held by the System by 1987 was made.

The Columbia, South Carolina Farm Credit System district reported foreclosure of over 1200 loans in 1984–1985. Farm prices are very low, but they will be lower if the Farm Credit System folds, a bank official suggested. Farmers' assets, especially land, are continuing to decline in value. Financial problems were reported in the Kansas City, Louisville, and Western Farm Credit System districts. The position across the entire System was that if farm problems continue, and this was expected, farm lenders can expect more and more financial stress.

A Chase Econometric Study commissioned by the Farm Credit Council in 1985 estimated that the holders of the System's $70 billion in bonds would lose $28 billion if the System defaulted. In addition, it estimated that the nation's gross national product would lose $32 billion and the deficit would rise $32 billion the first year. Such a default would have serious repercussions throughout the financial markets because investors would lose confidence in the bonds. Agriculture would lose its ability to go into the money markets for its funds.

WHAT SHOULD BE DONE?

In the face of the mounting evidence of the interrelationship between economic problems of the farm sector and the Farm Credit System and the tie-in of these problems with the whole national economy, farm and credit leaders, congressional leaders, and others began a search for possible solutions to the problem.

In late October 1985, Donald Wilkinson, Farm Credit Administration Governor, recommended to a House Agriculture Subcommittee that Congress establish a $5 billion backup line of credit for the Farm Credit System, to be drawn upon as needed.

He also recommended the Farm Credit Administration be granted expanded enforcement power and clarify its role as an arm's-length regulator of the whole system. In addition, Congress should mandate that the System be empowered to marshall all its resources across the entire System by establishing statutory authority for this purpose. He pointed out that the System is salvageable, but Congress needs to restore confidence in it. It will need help over the next 18 to 24 months as it may have to absorb over $13 billion in losses.

The position of the Congress was reflected by statements of various members. Senator Dole pointed out that he felt strongly that the System was in need of a thorough examination and hoped the Senate Agriculture Committee would be able to hold hearings on farm credit by the end of September. Bills were introduced in both Houses of Congress. Provisions of the bills differed somewhat, but there was a feeling on the part of members that something should and would be done, perhaps by the end of the year.

The position of the Administration was reflected by a statement of the Secretary of Agriculture that the Farm Credit System faces potential chaos unless some solution is found for the System by 1986. The Secretary of the Treasury and the president endorsed a position paper prepared by U.S. Department of Agriculture which called for regulatory and statutory changes in the System. The statement did not endorse a line of credit, but said the Administration would assess the need for federal financial assistance if Congress is willing to support those changes.

Despite the apparent general position that help was needed, there was an undercurrent of mild skepticism on the part of a few members of Congress. There was liberal usage of the term bail out and the connotations associated with it. There was also the view held by some that the System was a loose network of independent agencies consisting of the 12 Federal Land Banks, 12 Federal Intermediate Credit Banks, 13 Banks for Cooperatives, and, at the local level, several hundred Production Credit Associations. There was the implied position that the groups were interested in doing their own thing and that little, if any, coordination of the overall effort could be discerned.

In addition, there was the issue of extending help only to the Federal Farm Credit System and not to troubled commercial agricultural banks.

The position that it was in the public interest to help the System easily prevailed, however. Bills were passed by both Houses of Congress, the differences were speedily ironed out, and the legislation was presented to the president for his signature in late December. It was signed into law by the president a few days later.

THE AID BILL CREATES A NEW INSTITUTION

The aid bill created a new institution within the Farm Credit System which is called the Farm Credit System Capital Corporation. This institution will serve as a warehousing agency in that it will take over bad loans, so called nonperforming loans, in the System and centralize the surplus reserves of the System, at that time amounting to about $7 billion.

Once the reserves are exhausted, and the System officials felt that they would be within about 2 years at the most, the legislation authorizes the Secretary of the Treasury to provide funds to the Capital Corporation to keep the System operating. This action on the part of the Secretary of the Treasury can be taken, however, only if a separate appropriation bill is enacted by the Congress for this purpose. Credit System officials could return to the Congress as early as the end of 1986 with requests for such appropriations.

The Secretary of the Treasury would carry out this function, once it is approved in advance by an appropriations bill. Obligations issued by the Capital Corporation would be purchased under terms and conditions set by the Secretary. Prior to this authorization by the Congress, the Farm Credit Administration would have certified that the System's capital and reserves had been exhausted. Any further contribution of reserves by the institutions in the System would make it impossible for them to make loans to eligible borrowers.

The Capital Corporation will be administered by a five-member board of directors. Three members will be elected by the farm credit banks that own the voting stock in the Capital Corporation. Two members will be appointed by the Farm Credit Administration chairperson. If the Secretary of the Treasury provides funds to the Capital Corporation, two members would be added to the board. One of these would be appointed by the Secretary of Agriculture and one would be appointed by the other six board members. This appointee would have to come from outside the government and the credit system.

The authority of this Capital Corporation to redistribute funds and make assessments would expire on December 31, 1990.

ADMINISTRATIVE CHANGES

Another significant change was made relating to how the Farm Credit Administration is governed. The System will now be run by a three-member board of directors nominated by the president and confirmed by the Senate. They will serve 6-year terms on a staggered basis. They can serve only one term. One of the three members will be designated as chairperson by the president and will serve as chief executive officer. No more than two board members can be of the same political party. This board will replace the former 13-member board from the districts of the System. This former board will become an advisory group.

OTHER PROVISIONS

The Farm Credit Administration Board would set requirements for loan security and approve bond issues along with their interest rates. It would oversee and regulate the transfer of funds between institutions of the System. It would require annual independent audits of each institution and, at the discretion of the chairperson, carry out examinations of the institution in much the same manner as those carried out by commercial bank examiners. It would establish minimum levels of capital reserves for each institution in the System. It could require mergers of individual institutions. It is empowered to issue cease and desist orders against officers and institutions for violation of regulations, with power of removal of directors and officers. This role of the board is similar to that of the FDIC which regulates commercial banks.

IMPLICATIONS

Perhaps the most significant change made in the organizational arrangement of the Farm Credit System is in its governance. The relega-

FIG. 11.3. President Reagan signs the Farm Credit Amendments Act of 1985 as Vice President Bush, Secretary of Agriculture Block, and members of Congress look on.
White House photograph, courtesy of the Farm Credit Council.

tion of the 13 current members of the board to an advisory role could serve to remove the System from its agricultural administration. Whether this will, in fact, happen and what might be the ramifications of such a movement remain to be seen.

The seeming agreement on the part of the Congress, the Administration, and by implication, at least, of the public that such an institutional arrangement was needed to serve the unique needs of agriculture may be a most significant aspect emerging as a result of the System's economic crisis. The full extent to which the agreement was based upon an understanding of this uniqueness also remains to be seen.

The warehousing and cross-subsidizing role established for the new Capital Corporation appears very positive if used properly. The regulatory role of the new board, patterned after the FDIC, should also prove to be positive. It may well be that the Farm Credit System, based upon the "necessity is the mother of invention" adage, has been preserved and that it may have weathered adversity and emerged stronger and better able than ever to serve its role.

REFERENCES

Bank for Cooperatives 1979. Banks for Cooperatives, How They Operate. Circular 40.

Commodity News Service, Inc. 1985. 2100 W. 89th Street, Box 6053, Leawood, Kansas 66206.

Congressional Record. Proceedings and Debates of the 99th Congress, 1st session. Washington, DC.

Valko, L. 1981. Cooperative Laws in the U.S.A. Bulletin 0902. Washington State University, Pullman.

TO HELP IN LEARNING

1. Prepare for a 30-minute brown bag seminar with your peers on the topic: "The Unique Credit Needs of Agricultural Cooperatives." Cover the following: (a) sources of uniqueness, (b) examples of uniqueness, (c) use of usual sources of credit, such as commercial banks, and (d) specially designed credit sources.

2. Summarize your findings and position.

3. Is the requirement that initial funds advanced by government to start the Banks for Cooperatives be repaid sound? Why?

4. The bill passed by Congress to help the Farm Credit System weather the economic crisis in 1985 made it possible for the Secretary of the Treasury to purchase notes or obligations of the Credit System. In your judgment, what should be the responsibility of the Credit System for any losses suffered by the Treasury because of nonpayment of the notes? Why?

DIRECT QUESTIONS

1. Cooperatives could borrow from any or all of the usual sources of funds. Why don't they?

2. What is the basic underlying reason that usual sources of credit cannot very well be used?

3. How did the Farm Credit System get started?

4. Who provided the initial capital for the Farm Credit System?

5. Is the initial capital loan still outstanding?

6. Where do the Banks for Cooperatives get their loan funds?

7. Where are the Banks for Cooperatives located?

8. What are Production Credit Associations?

9. What are National Farm Loan Associations?

10. Where do agricultural cooperatives borrow their funds?

11. Are the Banks for Cooperatives cooperatives themselves?

12. What other services are provided by the Banks for Cooperatives?

13. What is the Farm Credit System Capital Corporation?

14. Why was the Capital Corporation formed and what can it do?

15. How was the Farm Credit System governed before 1985?

16. How is it governed after 1985?

17. How is the Capital Corporation board of directors selected?

18. What is a nonperforming note or obligation?

TYING-TOGETHER QUESTIONS

1. Build, block by block, a short position paper on the U.S. farm credit system and its rationale.

2. React to this statement, "Special credit arrangements for agriculture are not justified."

3. Compare the statements which you made in question (2) with those you made in question (1).

4. What is the relevance of seasonality in agricultural production to a Farm Credit System?

5. Comment on the statement, "Necessity is the mother of invention applies to the emergence of the Farm Credit System."

6. Prepare a detailed position paper on the subject, "Are there any differences between government aiding the Chrysler Corporation and its aiding the Farm Credit System?"

7. In your judgment, is the change in the governance of the Farm Credit System brought about by the 1985 credit bill highly significant? Explain your answer in detail.

12

Equity Redemption Plans
Used by Cooperatives

Our concern in the preceding chapters was with capital accumulation or formation in amounts to meet the requirements of our cooperative for operations and for longer-term capital needs. Plans were considered such as retained patronage earnings and member capital retains. All were considered in relation to the unique features of the cooperative corporation, especially from the standpoint of the implications of the member–user–owner trio in this unique institutional arrangement. A definite concern stemming from this uniqueness is with keeping investments by members in their cooperative in line with their use of the cooperative—the so-called rule of proportionality.

Hopefully, our cooperative has been able to devise capital accumulation plans of such a nature that its capital needs are adequately met and are in line with cooperative principles. We have accumulated member equity as shown by our accounting procedures, and each of our members has equity in the cooperative as shown by the accounts.

The other side of the member–user–owner trio concept is that not only must capital be accumulated to meet all capital needs of the cooperative, but at some stage, the capital accumulations or member investments must be distributed back to those to whom they belong. Their equity capital, which has been formed under the various plans, must be redeemed.

This chapter will concern itself with equity redemption plans which, again, preserve the integrity of the farm cooperative as being unique among business enterprises. Again, the member–user–owner concept will be dominant in our thinking as it relates to various plans of redeeming members' equity.

OBJECTIVES OF EQUITY REDEMPTION PLANS

Several alternative equity redemption plans will be discussed here, but perhaps criteria or objectives which cooperative leaders should have before them in choosing a plan or plans are appropriate at this stage. Such criteria include the following:

1. Any plan adopted should help in providing for the capital needs of the cooperative. These include cash flow for servicing debt, equity redemption, growth, and for making sure that services which members need and want are provided, both currently and in the future.
2. Any plan should facilitate conformation with the principle of proportionality, a basic cooperative position. Members should supply risk capital according to their use of the cooperative's services. Any redemption plan should enforce this requirement.
3. Be flexible. A wide range of membership needs, characteristics, and operating results must be accommodated. Potential impact of any equity redemption plan on relevant variables should be thoroughly assessed when alternative plans are being considered.
4. All discussions of alternative equity redemption plans and the final adoption of any plan(s) should take into account the management trio we have discussed, but everyone involved should recognize and accept the fact that the board of directors has the responsibility of controlling the cooperative's equity redemption policy. It must make all final judgments and accept responsibility for the effectiveness of any plan(s) adopted for use.
5. Any policy and any program designed to implement the policy should be (a) easily understood by members and employees who are in contact with members; (b) easily administered without excessive operating costs; and (c) completely compatible with all relevant tax laws, any other relevant laws, and relevant debt obligations of the cooperative.

While keeping these criteria and objectives in mind, let's examine a number of alternative equity redemption plans or programs.

THE REVOLVING FUND PLAN

A plan that is admirably adapted to both accumulation of capital and its redemptions is called the revolving fund. It is also known as the first in–first out or fifo plan for revolving capital, and when first conceived as a plan for redemption of equity capital on a chronological schedule, the "rotary fund." It is perhaps a term more closely associated with cooperative finance than any other. It came into being as a means of acquiring capital in accordance with a basic cooperative principle, a member's use of the cooperative. It has undergone changes and adaptations over time and is now a widely used plan for accumulating capital and for redeeming it on a chronological schedule.

Under this type of plan, a cooperative pays off in cash on a chronological schedule the oldest outstanding equities when a net worth position that is satisfactory has been achieved. This plan, if properly designed and implemented, obtains funds for capital uses from members in proportion to their use of the cooperative, through retained patronage earnings or per unit capital retains and retires, or pays them off on a planned chronological schedule. It adjusts ownership of the cooperative in accordance with cooperative principles through this double process of accumulating capital and redeeming it.

HOW A REVOLVING FUND WORKS

An example of how a 5-year revolving fund for a cooperative member would work is shown in Table 12.1. The accounting procedure illustrated in the example is straightforward and assumes that the new equity for the member, beginning in 1980, was $600 and came from the member's paid-in initial capital when the cooperative was started in 1980. New equity is earned each year by patronage earnings, the share of this member's net savings, or net income realized by the cooperative being retained and allocated to the member. New equity might also be earned by the cooperative retaining a certain amount or a certain percentage of the receipts from sales of the product sold for the member by the cooperative. These funds are accumulated in the member's account by the cooperative and are shown in the total column.

TABLE 12.1
An Example of a 5-Year Revolving Fund for a Cooperative Member

Year	Beginning balance[a]	New equity[b]	Total	Amount redeemed[c]	Ending balance[d]
1980	0	600	600	0	600
1981	600	700	1300	0	1300
1982	1300	800	2100	0	2100
1983	2100	600	2700	0	2700
1984	2700	700	3400	0	3400
1985	3400	800	4200	600	3600
1986	3600	600	4200	700	3500
1987	3500	500	4000	800	3200

[a]All figures represent dollars. Assume paid-in initial capital.
[b]From retained patronage earnings or from per unit capital retains.
[c]Amount of new equity 5 years before.
[d]Member's net allocated equity as shown on the cooperative's books.

Beginning in 1985, the cooperative will start revolving the fund by redeeming, in the fifth year, the new equity which was allocated to the member in 1980. The new equity earned in 1981, $700, will be revolved or redeemed in 1986.

This is a simple illustration of how a revolving fund accumulates capital and systematically redeems it on a planned chronological schedule. It is obvious that many plans based upon this concept may be much more complex. They could, for example, be designed to mix features of other plans, mix ways of accumulating equity, and for its redemption segregate funds by divisions, by commodities, and income from other sources such as other cooperatives and nonmember business. Quite often, some sort of special arrangement for redemption of equity in case of a member's death or other event affecting patronage is incorporated into the plan. The longer the revolving period, the greater the need for other special features. With a short revolving period, there is less need for special features.

The ideal length of a revolving period is difficult to determine under all circumstances, but the overriding concern is with maintaining a healthy balance between accumulating necessary equity capital and its redemption or payout.

Short revolving periods serve the important cooperative principle relating to currency, that is, keeping investment in the hands of those who are currently using the cooperative. The need for special plans for redeeming equities of estates of deceased members out of sequence is reduced or eliminated by the use of short revolving periods. However, they require relatively large net margins, large per unit capital retains, or some other source of capital if other demands on cash flow are to be met.

Long revolving periods are often used by cooperatives to meet their ever-increasing needs for capital. Such periods, however, are unsatisfactory for a number of reasons. They wreak havoc with the currency principle. They create problems for inactive members and with payments to the estates of deceased members. Ownership patterns are distorted and many of the cooperative principles are violated. They may cause questions to be raised by taxing authorities. They may be used as a seemingly easy way to meet increasing capital needs and preclude the use of analytical procedures in determining long-term analysis of the cooperative's situation and the launching of a short- and longer-term plan of operation designed to place and keep the cooperative on sound footing. This calls for well-conceived plans for both capital accumulation and its redemption.

PROS AND CONS OF
REVOLVING FUND PLANS

Basic reasons for the revolving fund becoming the most popular equity redemption plan used by cooperatives include the following:

1. This type of plan is simple and easy to understand and administer. Cooperative members understand it best. It is flexible and adaptable for almost any situation or need.
2. It does not violate cooperative principles relating to proportionality, currency, and so on when properly designed and implemented.
3. This plan is subject to midcourse adjustment by the board of directors to ride through economically bad years or other circumstances, provided knowledge regarding the cooperative's situation is adequate and provided the cooperative's bylaws permit adjustments by the board of directors as they deem necessary.

As usual, no plan or system is without possible shortcomings and revolving fund plans are no exception. Some of the plan's drawbacks are suggested in the following:

1. The length of the revolving period can be easily extended. This degree of flexibility may be a positive factor in most cases, but it may also be used to cover the results of poor planning, poor judgment,

and generally poor operating results. It may be used to rationalize a lack of planning.
2. Unless adjustment procedures are provided, differences between benefits received from the cooperative and investment made in the cooperative may come about. This could happen when margins earned by the cooperative vary significantly over time and if the fund covers several products or services whose margins vary.
3. Lacking proper member education programs, members may develop unrealistic expectations regarding redemption of their equities regardless of the cooperative's financial condition. Perhaps this is not really a shortcoming of the fund itself, but of a lack of complementary action on the part of the cooperatives.

While no plan or system can claim to represent a financial panacea, the revolving fund plan has much to offer. Its ability to be tailored to fit most circumstances, its straightforwardness, its ease of administration, and its potential for operating in conformity with basic cooperative principles go far in recommending its use.

When a well-designed plan is developed and implemented and its use is supplemented with member education programs about the plan, the chances of it operating successfully are great. Most of the criteria set forth in the beginning of this chapter are met by this type of plan.

THE BASE CAPITAL PLAN

The underlying rationale for a base capital plan for financing a cooperative rests on the position that a cooperative should determine its required amount of equity and systematically adjust this base amount to fit current needs and members' use of the cooperative. A member's equity contribution to the cooperative would be tied to use of the cooperative, a basic cooperative principle. Also, the equity needs of the cooperative would be systematically determined and proportionately distributed to the members.

A cooperative, in adopting a base capital plan for capital formation (it is also a capital redemption plan), would take these steps:

1. Determine a base amount of capital which the cooperative needs.
2. Determine an appropriate base period, and each member's share of the total base amount needed would be based on proportional use of the cooperative during that period.
3. The total base capital needs of the cooperative would be adjusted

systematically, usually annually, to reflect current needs of the cooperative and each member's contribution to the total adjusted accordingly.

4. Those members who are underinvested, that is, their use of the cooperative is greater than their contribution to the total base capital needs, would have to add to their contribution in accordance with a systematic plan.

5. Those members who are overinvested, that is, contributing more than their proportionate share to the total needs, would be paid the amount of the overage in cash, also in accordance with a systematic plan.

The proportionality principle is well served by use of a base capital plan, since it relates a member's investment to the current proportional use made of the cooperative. If member X markets twice as much milk through the cooperative as does member Y, then member X's investment in the cooperative should be twice as much as that of member Y. Also, if the plan is soundly developed and correctly reflects the amount of equity capital needed by the cooperative, the needs of the cooperative will be adequately served. Couple this with forecasting the needs of the cooperative on the basis of members' proportional use of the cooperative and the result is an adequately financed cooperative with the principle of proportionality fully operative. Such a plan usually includes a systematic way of returning equity to those who have overinvested and for redeeming the equity of those who are no longer active members of the cooperative.

If a board of directors should decide to use a base capital plan, the following steps would usually be taken:

1. Determine how much total capital will be needed by the cooperative for the coming fiscal year and for long-term programs and how much should come from internal sources and be equity or net worth.

2. Decide the type and amount of allocated equity that will be designated as base capital. Usually, capital from direct investments, per unit capital retains and/or retained patronage refunds, are designated as base capital. Capital in the form of preferred stock, unallocated reserves, and so on is not included in the base capital.

3. Once the cooperative's total capital needs are known and which capital is to be included in base capital, each member's proportionate share of the base capital can be determined. This is determined by the proportion of the cooperative's business which the member did with it during the base period used. Usually the past 2 to 6 years would be used as the base.

4. Using current equity capital holdings by each of the members, a determination can be made as to whether a member is overinvested or underinvested.

5. A method would then be devised for accumulating the needs of capital from those who are underinvested and for redeeming the appropriate amount of capital for those who are overinvested.

An example of determining a member's proportionate share of the needed total base capital of a cooperative is shown in Table 12.2. Assumptions used in constructing the table are that the cooperative has five members and $50,000 of equities at the end of its fiscal year and needs $60,000 of base capital for the next fiscal year. Using the past 3 years as the base period, each of the five members' proportionate share of the needed $60,000 of base capital was determined. This proportionate share of the needed capital was then matched with the member's current equity capital to determine whether the member was over- or underinvested. There is, of course, the third position—that of a member having a current account that matched the proportionate share of the needed base capital. As shown in the table, members A and D were underinvested, members E and C were overinvested, and member B's current and required investment were exactly matched.

Table 12.2 also shows the adjustments necessary to bring members' equities into proportionate alignment with patronage. If a particular member is substantially underinvested, the full adjustment necessary to match current and required base in one fiscal year may not be feasible. This would likely be true with new or younger members with possible problems with cash flow. The board of directors would need to develop policies in regard to this and similar situations. On the other hand, the cooperative might not be able to redeem in full the equity of members who are heavily overinvested in one year. Again, a definite cooperative policy is suggested.

LENGTH OF BASE PERIOD
IN BASE CAPITAL PLAN

One of the important decisions to be made by the board of directors in implementing a base capital plan relates to the number of years of marketings or history which will be used in determining each member's patronage base period. A short period of time, such as 2 to 4 years, will reflect changes in price structure more quickly than would longer periods, say up to 10 years. Prices in the base period and current period could, of course, be adjusted for inflation by an appropriate factor when

TABLE 12.2
Example of Calculations Necessary in Setting Up a Base Capital Plan

| | | Allocated equity capital (thousand dollars) | | | | | |
| | | End of fiscal period | | | | Next fiscal period | |
Member	Share of needed base capital (%)[a]	Current base[b]	Required base[c]	Over or under[d]	Retain or refund added[e]	Equity redeemed[f]	Current base[g]
A	20	10	12	-2	2	0	12
B	25	15	15	0	0	0	15
C	15	10	9	1	0	1	9
D	30	10	18	-8	8	0	18
E	10	5	6	1	0	1	16
Total	100	50	60	-8	10	2	60

[a]Percentage that the member's patronage was of the cooperative's total patronage by all members during the base period—the previous 3 years.

[b]The amount of allocated equity the member has with the cooperative at the end of the fiscal period and the total for all members.

[c]The amount of equity capital required of each member based upon the member's patronage during the base period. Required base times percentage shown for share of needed base capital.

[d]Current base minus required base.

[e]If underinvested, a retain is added above current base; if overinvested or if current and required bases are the same, nothing is added.

[f]If a member is overinvested, equity is redeemed—current base less required base.

[g]Member base capital and total cooperative base capital.

longer periods are used. A shorter base period will keep investments more in line with changes in membership and patronage and will minimize or eliminate the need for redeeming estates out of sequence.

A long period would even out fluctuations in a member's required base or in capital requirements, but would not be reflective of any drastic changes that might come about in the cooperative due, perhaps, to drastic changes taking place in the industry with which the cooperative deals. Perhaps some sort of average of a reasonable number of years is suggested.

A soundly conceived length of base period would take into account the cooperative's ability to generate net savings or capital retains. This is especially true, of course, if the cooperative depends upon allocated net margins as its source of financing. The level of new investments may not increase enough to redeem the equities of the members if the cooperative does not realize sufficient savings or net margins. It is because of this potential source of instability in financing that cooperatives use per unit retains rather than allocated margins in their base capital plans.

UNIT FOR CALCULATING EQUITY

Another decision the board of directors will need to make is the unit the cooperative will use in determining each member's patronage. In many cases, the unit used is physical in nature, for example, a bushel or hundredweight (cwt), a ton, or a gallon. Problems arise, however, in using physical units to measure the proportionate use of the cooperative by each member and to determine the base capital requirements of the member in the cooperative.

One problem with using physical units for these purposes is that it does not reflect other services provided by the cooperative and changes in the value of the unit. A certain number of cents per hundredweight of milk handled might be sufficient when milk is $5.00 per cwt, but when the price rises to $14.00 per cwt, the cooperative becomes undercapitalized even though it is receiving the same number of hundredweights as it had before. This problem has led to the use of indexing the per unit retains in accordance with some appropriate price level measure. Of course, the dollar value of the units could be used as a measure rather than just the number of physical units.

OTHER DECISION AREAS

Other decisions cooperative boards of directors will need to face include setting priorities for retirement of equities, limiting equity investments and redemptions, how overinvested and underinvested members will be brought into proportionality, and how retired members and deceased members' estates will be handled. Let's examine each of these areas as a part of establishing a base capital plan.

In the usual case, many members, usually the newest and probably the youngest ones, have relatively little equity in their cooperatives. Requiring them to build up their capital to complete proportionality in a short period of time may cause cash flow problems and undue financial hardships. Careful study of the situation may indicate the need for a limit on the amount the members are required to provide in a single year. For example, some cooperatives require new members to provide 15% of the value of their products. Other cooperatives may require a new member to pay a minimum of 10% of the member's base capital requirement. Establishing a definite period of time, say 5 to 8 years, in which a member is required to make the full base capital investment requirement is another plan that lends itself to orderly handling of the cooperative's base capital requirements. It also takes into account possible financial difficulties new members may have in building proportionate equity in their cooperative.

The other side of the coin, of course, is the equity redemption schedule of the cooperative. The cooperative simply may not have sufficient funds to redeem all excess equities in 1 year without jeopardizing its equity position. In such cases, again after careful study, the board of directors may establish a policy of limiting such redemption. In one case, a cooperative redeems 12.5% of the total each year, thus spreading the redemption of such equities over 8 years. Others establish limitations on equity redemptions to some specified minimum of funds, sometimes tied to a certain percentage of any gain in net worth which the cooperative might experience over an accounting period.

Decisions regarding equity building and redemptions are made by the cooperative board of directors, of course, because they establish cooperative policy in all cases and are responsible for the cooperative's financial health. Prudent judgment that reflects the cooperative's circumstances at a given time is essential. Flexibility in the programs designed to carry out the policy is necessary in order to take into account the cooperative's situation. Complete understanding of the cooperative and information based upon proper analytical procedures are essential if sound decisions are to be made by the board of directors regarding level of capitalization, unit of measurement, base period for use in determining the volume of business, and equity retirement priorities. With sufficient data and information and with creative planning, all relevant items can be incorporated into a workable plan which serves the cooperative's need for equity capital and makes it possible for each member to contribute equitably to those needs.

Some cooperatives use what is referred to as the "most overinvested member" plan in retiring capital after the cash patronage refund has been paid. A determination is made of the percentage that funds available for equity redemption would allow, and this amount would be applied to redemption of the equity of those members who had been deemed by the board to be most overinvested. Of course, those in underinvested positions would continue to accumulate deferred refunds on capital retains and would not share in any equity redemptions until they reached the required level.

A variation of the most overinvested plan which permits a wider sharing of equity redemption by more members is the "all overinvested member" approach. In this case, as with the previous case, a determination would be made by the board as to how much funds are available for retiring of capital. Then a portion of this amount would be paid to each member who was overinvested in accordance with the percentage of overinvestment. Again, those who are underinvested or who are in a balanced position with respect to base capital requirements would not share in the equity redemption.

Plans for redeeming the equity of retired members and for payments to estates of decreased members are necessary if the cooperative is to maintain the principle of currency (i.e., those who are currently using the cooperative should finance it). There may also be laws relating to equity redemption under these circumstances. In some cases, equities are due for payment over a period of time as determined by the board. This lends itself to a degree of orderliness, since expectations of the number of members in these categories and amounts of equity involved can be actuarially determined on the basis of the cooperative's makeup and experience, and funds can be established for this purpose. Some boards have decided that estates will be paid before any other equity retirement. Other plans may be used, but the important aspect is the recognition of the need to plan for such capital redemption by the board and to take systematic steps in handling this area.

ADVANTAGES AND DISADVANTAGES
OF BASE CAPITAL PLANS

No plan for accumulating and redeeming equity capital can be said to be best for all cooperatives. The objectives of a cooperative with respect to capital needs, some degree of sufficiency, and in keeping with accepted principles of proportionality, currency, and the like are rather standard, but the method or methods used to achieve those objectives probably vary from cooperative to cooperative. This means that careful study is an essential aspect of adopting any plan and tailoring it to fit the cooperative's unique circumstances at a given time. Let's now consider some of the advantages and disadvantages of a base capital plan in this light. Some of the advantages of this type of plan are as follows:

1. It serves the principle of currency very well because it directly ties the member's share of equity capital in the cooperative to use of the cooperative. Furthermore, it provides an easily understood and systematic method of adjusting investment to use and maintaining an equitable ownership structure.
2. It allows an orderly transfer of cooperative ownership from past users to current users.
3. It is adaptable and flexible in that it permits a cooperative to fit a plan to its particular needs and to change the plan and its provisions as circumstances change. The plan can be expanded or contracted as capital needs expand or contract.
4. It is flexible and adaptable in regard to retirement of capital, with the needs of the cooperative serving as guides for the flexibility.
5. Emphasis is placed on the member's equity accumulation in the cooperative as representing an investment rather than funds being withheld from the member and not available for use. This is in keeping with the service–user–owner relationship which is a distinctive feature of the cooperative corporation.
6. This type of plan, perhaps more than any other, permits and requires more accuracy in budgeting, cash flow estimation and long-term planning and so on because capital sources are more predictable and reliable.
7. Cash refunds to members are more likely under this type of plan since it is grounded on the basis of systematic capital accumulation and redemption.
8. It may engender more interest and more involvement of members in their cooperative.

As in all plans, there are also disadvantages in cooperatives' use of a base capital plan. These include the following:

1. It may cause undue financial hardships because of lowered cash flow for new members, usually young farmers, if established with inflexible provisions.
2. If membership is unstable and has a high turnover rate, for any reason, such a plan will not work well.
3. Some boards of directors are inclined to take an easier way out in

meeting the cooperative's increased capital needs and lengthen the revolving fund period rather than asking members for increased contributions to the base capital plan.

4. An appropriate length of base period for such plans is difficult to establish. Base periods that are too short may give rise to financial problems for members and then to the cooperative if environmental or other conditions result in a bad crop year and certainly if several such years occur in succession. If the period is too long, problems will arise in meeting the currency principle, that of keeping the obligation for providing capital to the cooperative in the hands of those who are currently using it.

5. Concerted educational efforts for members are essential if such plans are to be effective, since they are generally considered to be complex and hard to understand.

6. Constant, critical examination of such plans in relation to the cooperative's capital needs is essential. Reluctance on the part of a board of directors to do whatever is necessary to maintain this position of knowledge may result in what appears to be a capital plan which is not working.

A close examination of the items listed as disadvantages readily invites questions as to whether the disadvantages lie with the plans themselves or with their implementation. Success of the plans is dependent upon having rather complete and current knowledge regarding the cooperative's capital needs and the capability of a base capital plan to meet those needs under varying situations. In addition, member education and development of member interest and involvement are a part of the necessary and sufficient conditions for the smooth working of such plans. This suggests that given the commitment to knowledge and understanding and to membership education, plans can be developed which have appropriate length of periods and the required flexibility and adaptability. They can be tailored to fit the needs of the cooperative and to take into account the peculiar financial needs of members. Those things necessary to make a base capital plan work effectively are some of the very same things necessary for effective cooperative performance in general. The suggestion made previously may be valid that the areas mentioned as being disadvantages may not in fact be shortcomings of the concept, but rather of the conditions that have a major bearing on its success. The advantages of the base capital plan for accumulating and redeeming member-allocated capital appear to far outweigh the so-called disadvantages.

PERCENTAGE OF
ALL EQUITIES REDEMPTION

A very small percentage of all farmer cooperatives use a plan in which the board of directors determines some percentage of all allocated equity that will be redeemed during a period, and this percentage is used to calculate the amount redeemed for each member. If, for example, the board of directors determined that the cooperative had $50,000 from net savings that could be used for redeeming allocated equity of $500,000, this would mean that 10% of the equity could be redeemed. Then 10% of each member's allocated equity as carried on the cooperative's books would be redeemed in cash regardless of date of issue or share of equity held by the member. An example of such a plan is shown in Table 12.3.

In the illustration, the board of directors determined that their net savings would support a 5% redemption of all members' allocated equity. Their policy was that the amount to be redeemed would be based on the beginning equity balance. This member was paid nothing in the first year because of the zero beginning balance. Of the beginning balance in the second year, 6%, or $30, was redeemed, and a balance of $1070 was left in the member's allocated equity account. A member who becomes inactive would continue to receive whatever percentage of total equity

TABLE 12.3
How a Percentage of All Equities Plan of Redemption Would Work for a Member

Year	Beginning balance ($)[a]	New equity ($)[b]	Redemption[c] Percentage	Amount ($)	Ending balance ($)[d]
1	0	500	5	0	500
2	500	600	6	30	1,070
3	1,070	500	10	107	1,463
4	—	—	—	—	—
5	—	—	—	—	—
6	—	—	—	—	—

[a]Assume paid-in capital.
[b]From retained patronage earnings or from per unit capital retains.
[c]Percentage and amount of beginning balance to be redeemed.
[d]Member's net allocated equity as shown on the cooperative's books.

that the board determined was available until the account was completely retired or some event, covered in the cooperative policy, brought about a different payment schedule—perhaps a balance or balloon payment at the death of the member.

Again, there are advantages and disadvantages to this type of plan. Advantages would include the following:

1. This type of plan is not complex—it is easily understood.
2. New members do not have to wait as long under this plan to receive redemptions as under most other plans.
3. It is easy to administer.
4. It is adjustable to different operating results based on year's net savings and capital requirements of the cooperative.

Shortcomings of this type of plan are as follows:

1. The redemption process in the plan slows the process of shifting ownership from overinvested and inactive members because redemption is spread across the board to include all equities.
2. It may have to be combined with other plans to handle inactive member redemptions or other special cases.
3. In those cases where cooperative membership is mobile and patronage is erratic, the plan would not be feasible.

A SPECIAL PLAN
FOR EQUITY REDEMPTION

The term plan in this case may be a misnomer, since it is not really planned but is triggered by some unpredictable event such as a member's death, a member leaving farming, member retirement, special hardship such as bankruptcy, or a member moving away from the area in which the cooperative provides services. More than one-third of all cooperatives, however, are reported to use only this type of equity redemption procedure, and another 20% use it as part of another plan.

In a special plan, equity is accumulated and held as allocated equity in the accounts of members until the condition or event that has been established by the board as a basis for equity redemption has been met. When the prescribed event or condition has been verified as having occurred, the administrative work necessary to make the redemption is set in motion and the entire amount of the equity is redeemed in one lump sum or over a prescribed period of years if that is the policy established by the board. If cooperative policy permits, the board may set a maximum limit that may be redeemed in a year, and it may also establish priorities for equity redemption based on the events that may occur to trigger the redemption process. Prioritizing would be essential when funds available for redemption of equities are limited. They may be redeemed in cash or, at the discretion of the board, be converted to interest-bearing securities which may be at face or a discounted value.

An illustration of a special plan for equity redemption is shown in Table 12.4. In this example, equity is accumulated and no redemption takes place until a specified condition is met, perhaps in the case of the death of a member. After 3 years, the entire amount of allocated equity is paid to the member's estate.

The suggestion was made at the beginning that this system might be referred to as a "planless plan," since for the most part it is tied in with unpredictable events. Some have suggested that it was a system designed to avoid planning and for reactions to events rather than anticipating them in some sort of systematic way and being prepared to act in an orderly fashion when they occur.

This statement or allegation may be a bit harsh because the weakness in this plan which might lead to a lack of financial planning may be minimized by taking certain steps. Records of the probability of occur-

TABLE 12.4
An Example of a Special Condition Equity Redemption Plan

Year	Beginning equity balance[a]	New equity[b]	Equity redeemed[c]	Ending equity balance[d]
1	0	2	0	2
2	2	3	0	5
3	5	2	0	7
4	7	3	0	10
•				
•				
•				
20	30	5	0	35
21	35	0	0	35
22	35	0	0	35
23	35	0	35	0

[a]All figures represent dollars. Assume paid-in capital.
[b]From retained patronage earnings or from per unit capital retains.
[c]The prescribed event occurred in year 21 and no new equity earned afterward.
[d]Entire amount of allocated equity paid to member's estate after 3 years.

rence of each condition that was prescribed along with the corresponding equities involved could be compiled. With these data and related information, definite plans can be made to provide the necessary cash flow.

An equity profile showing the equity owned by different age groups would be relatively easy to compile. A mortality rate schedule with the member's age tied to the amount of equity owned would provide the basis for calculating probable equity redemption requirements. These could be incorporated into some financial planning horizon, say, a 5-year period, thus moving this event from one to which the cooperative reacts in a hit-or-miss fashion to one for which it specifically plans. Developing probabilities for other prescribed events such as members going out of business, moving away, or going bankrupt would not be beyond the realm of possibility, especially after the cooperative had developed experience over a number of years in a planned, purposeful manner.

Let's now examine some of the advantages and disadvantages of this arrangement for equity redemption. Advantages include the following:

1. The equity redemption burden on cooperatives is relatively light, since only specific conditions trigger equity redemption.
2. The plan is simple and easy to understand.
3. It is flexible, adaptable, and relatively easy to administer.

Disadvantages include the following:

1. It is not in compliance with the currency and proportionality principles.
2. Events that may trigger redemption, with the exception of age, are unpredictable and complicate the planning process of the cooperative.
3. Some of the events may be within the control of the member and triggered for the benefit of the member at the expense of the cooperative. Moving away from the area is an example.

Perhaps it is correct to say what has become obvious to the reader, that no plan for member equity accumulation and redemption is without shortcomings and no plan is perfect for any and all situations.

This places the burden for developing, adapting, and adopting a plan for a particular cooperative exactly where it should be—in the hands of the board of directors and the manager. With complete and current knowledge regarding the cooperative's capital needs, knowledge of relevant characteristics of its services and marketing functions, and with sympathetic and knowledgeable understanding of the members served and the importance of their interest in and knowledgeable involvement in their cooperative, there is reason to believe that an appropriate capital accumulation and equity redemption plan for a cooperative can be developed.

REFERENCES

Brown, P. F., and Volkin, D. 1977. Equity Redemption Practices of Agricultural Cooperatives. FCS Research Report *41*. U.S. Department of Agriculture.

Cobia, D. W. 1980. Alternative Equity Redemption Policies and Programs and the Need for Change. NICE, University Park, PA.

Dahl, W. A., Dobson, W. D., and Veium, D. H. 1981. Mandatory Equity Retirement Plans for Farmer Cooperatives. Economic Issues, No. *55*. University of Wisconsin, Madison.

Harling, J. A. 1980. Equity Retirement by Cooperatives—Positive Reaction to Legislative Concerns. NICE, University Park, PA.

Royer, J. S. Financial Aspects of Equity Redemption. ESCS, U.S. Department of Agriculture.

U.S. Department of Agriculture, ACS. 1982. Equity Redemption, Issues, and Alternatives for Farmer Cooperatives. ACS Research Report No. *23*.

U.S. Department of Agriculture, FCS. 1983. Equity Redemption Guide. Cooperative Information Report No. *31*.

TO HELP IN LEARNING

1. It has been said that there's really no difference in payout plans used by agricultural cooperatives and dividends paid by private corporations. Comment professionally.

2. Adherence to the rule of proportionality is a goal in agricultural cooperatives. Is this goal also achieved in private corporations? Explain fully.

3. Mention the elements of a redemption plan which are tied in with the member–user–owner concept and whatever rules that may have stemmed from it.

4. Conduct a bag lunch seminar with your peers on the strategic and unique characteristics of agricultural equity redemption plans.

5. Conduct a similar seminar on the strategic and unique characteristics of payout plans of private corporations.

DIRECT QUESTIONS

1. Why is equity redemption itself so important for agricultural cooperatives?

2. Why are equity redemption plans themselves so important for agricultural cooperatives?

3. Which is more important for cooperatives, capital accumulation plans or equity redemption plans? Explain your answer.

4. What is a revolving or rotary fund plan?

5. What are the strong points of a revolving fund plan? Its weak points?

6. How long should the revolving period be? Explain.

7. Compare a base capital plan with a typical revolving fund plan.

8. What are the strong points of a base capital plan? Its weak points?

9. Why does the rule or law of proportionality keep coming up in considering almost everything about agricultural cooperatives?

10. What is considered in determining the length of time used for a base period in a base capital plan?

11. What are the factors considered in determining the unit that will be used in determining each member's patronage in a base capital plan?

12. Why is a policy regarding limiting equity redemption payments at times necessary?

13. Who is a most overinvested and most underinvested member? Why might this situation arise?

14. Why is a plan for redeeming the equity of retired or deceased members important?

15. The best plan, not properly implemented, becomes a poor plan. Comment and use examples.

16. What are the pros and cons of a percentage of all equities plan of equity redemption?

17. Why are special plans for equity redemption sometimes referred to as not really being a plan?

18. What are the advantages and disadvantages of special plans?

19. What is the special role of the board of directors and management in the area of equity redemption plans?

20. Comment on the role of members' education in the area of equity redemption plans.

TYING-TOGETHER QUESTIONS

1. It has been said that soundly conceived and implemented capital accumulation plans and equity redemption plans are the ultimate in cooperation. Drawing upon all your knowledge of the what, why, and how of cooperatives, comment fully.

2. Under what conditions in a cooperative would you say such plans would be developed and effectively implemented?

13

Cooperatives—
Taxation and the Law

Our objectives in this chapter are aimed at covering questions relating to taxation of agricultural cooperatives. This is an area that is frequently controversial and in which the cooperative is often attacked. The effort of the National Tax Equality Association is directed at what they argue is an area of unfairness by claiming that proprietary corporations are subject to double taxation while cooperative corporations' taxing is legally handled in such a way that the cooperative pays the tax or patrons pay the tax—a single tax basis. In the case of the private corporation, the corporation pays taxes on its net margins and the stockholders pay taxes upon any dividends paid to them.

This argument refers to federal income taxes and relates to areas where cooperatives differ from certain other forms of business organization. It is within these areas that controversies have sometimes raged. Because different positions often arise regarding federal income taxes most of our effort in this chapter will be devoted to them despite the fact that cooperatives, just as in the case of ordinary business corporations, are subject to many taxes.

TAXES PAID

As indicated, cooperatives are subject to many kinds of taxes, as with other forms of business corporations. They pay real property or real estate taxes, sales and use taxes, excise taxes, franchise taxes, and so on. In addition, they are subject to state and federal income tax laws. Tax treatment in this area is often contrary to the thinking in many sectors of public opinion and, as indicated, the area where controversies arise.

WHAT IS INVOLVED IN THE ARGUMENT?

First, let's review some of the basic elements of cooperatives and recall our developing an economic rationale for their existence. Let's recall the discussion and arguments that led up to passage of the Sherman Antitrust Act, the Clayton Act, and the problems encountered by agriculture with these Acts. Then we recall that the Capper–Volstead Act was passed in 1922 which made it legal for farmers to band together to market their products cooperatively without, per se, being in violation of antitrust legislation.

Further, let's brush up on our knowledge regarding cooperatives which we've gained so far by considering some definitions. This may help us in studying the tax issue and perhaps to reach a position we think is sound. Let's consider these areas. They will be given in summary form and will then be elaborated upon.

1. What is a cooperative? A cooperative is a business organized or formed for the purpose of providing goods and services for its patrons or marketing their products. It complies with the basic federal enabling legislation, the Capper–Volstead Act, and with cooperative legislation of the state in which it is incorporated.

2. Who is a patron? A patron of a cooperative is a person, firm, corporation, or association with whom or for whom a cooperative does business on a cooperative basis, whether a member, nonmember, or shareholder of the cooperative.

3. How are Capper–Volstead cooperatives taxed? Cooperatives are taxed under the provisions of corporate income tax statutes, under the single tax concept in which the cooperative pays the tax or the patrons pay the tax.

4. What are patronage dividends? A patronage dividend is an amount paid to a patron by a cooperative on the basis of quantity or value of business done with or for such patrons. The obligation of the cooperative to pay existed before the cooperative received the amount so paid. The amount paid is determined by reference to the net margin of the cooperative from business done with or for its patrons.

5. *What are per unit retains?* Per unit retains are allocations made by the cooperative to a patron on the basis of the quantity or value of the products marketed. The amount is not fixed in relationship to the net margin of the cooperative.

6. *May patronage dividends be qualified for exclusion from income tax obligation of the cooperative?* Yes, this can be done provided the patron is given a written notice of such exclusion and at least 20% of the dividend is paid to the patron in cash or qualified check. The patron must agree to include the dividend in income by writing to the cooperative, by consent in bylaws in cases where the bylaws of a cooperative require including the dividend as income as a requirement to become a member, or by qualified check where the patron by endorsing the check agrees to a statement to include the entire dividend, including the noncash portion, as income. There is also the requirement that the allocation must be made within $8\frac{1}{2}$ months following the close of the cooperative's fiscal year.

7. *What is a qualified written notice of allocation which a cooperative may make to its patrons to qualify for exclusion from income tax obligation?* This would include a notification of the allocation the patron can redeem in cash at face value or a notice in which the patron has consented to include the amount in taxable income in the same manner as cash. Such notices could be in the form of capital stock, revolving fund certificate, retain certificate, certificate of indebtedness, letter of advice, or other applicable written notice.

8. *What are the requirements to qualify per unit retains for exclusion from being taxed as income by the cooperative?* Again, a written notice of the allocation must be made and the patron must consent to include the per unit retain in income. This consent takes the same form as in the case of dividends. The allocations must be fixed without reference to the net margin of the cooperative.

9. *What are Subchapter T and Section 521 of the Internal Revenue Code?* These are the basic statutory provisions which center on federal income tax laws relating to cooperatives.

10. *What are the requirements for a cooperative to qualify for exemption from taxation under Section 521 from income tax obligation at the cooperative level?* It must be a farmer, fruit grower, or a like association. It must be organized on a cooperative basis to market the products of members and other producers or to purchase supplies and equipment for the use of its members or other persons. If it is organized on a capital stock basis,

substantially all its stock other than preferred, nonvoting stock must be owned by producers who are marketing products or purchasing supplies through it. Also, the dividend rate paid on capital shares must not exceed the greater of the legal rate of interest in the state where it is incorporated or 8% a year, whichever is greater. The financial reserves of the cooperative cannot be larger than those required by state laws or those that are reasonable and necessary. Trade or transactions with nonmembers may not be greater than with members. In the case of a purchasing cooperative, trade with nonmembers or nonproducers must not exceed 15% of the value of all trade. The cooperative must treat nonmembers the same as members in all business transactions and dividends. A permanent record of the patronage and equity interests of all members and nonmembers must be maintained.

11. How are earnings from nonpatronage business such as rental income, interest, and government handled under Section 521? They are allocated to members on the basis of their patronage. They are subject to the same requirements as patronage dividends for exclusion from taxation at the cooperative level.

12. What income, if any, remains as a residual and is taxed at the cooperative level? Assuming all Section 521 requirements are met, income from nonproducer, nonmember business is taxed at the cooperative level. In the case of federated cooperatives, the "look through" principle is used in determining exclusions from taxation.

HOW DID ALL THIS EVOLVE
AND WHAT'S THE RATIONALE?

The statutory references were quite limited until 1951. The basic rules of taxation of cooperatives were developed by the U.S. Treasury Department and in the courts. Two fundamental principles had evolved by that time, however, which have been considered as basic considerations in taxation of cooperatives. These were (1) exclusion at cooperative levels, and (2) taxable to the patron.

It was reasoned that a cooperative that distributes its net savings or margins to its patrons in proportion to their patronage based upon an obligation to do so which existed at the time the patronage occurred, in the form of bylaws, and so on, was exempt and therefore entitled to exclude those net savings from its income. The reasoning was that the cooperative had not realized income because whatever amounts that were involved belonged to the member–patrons from the beginning. By the same line of reasoning, if income had been realized, it belonged to the patron and was distributed or allocated to the patron. It should then be taxed at that, the patron, level.

These two principles have been referred to as the price adjustment theory and, at times, the single tax concept. Of course, both concepts are based upon the presupposition that the cooperative is doing business on an at-cost basis for its patrons and that it has no income itself.

Cooperative tax laws underwent a major rewriting in 1951. The Revenue Act of 1951 accepted these two concepts by providing for exclusion from taxable income a cooperative's patronage dividends or refunds made under a prior mandatory obligation to patrons. In addition, certain other exclusions were permitted so-called exempt cooperatives if the cooperative took the necessary steps that were set forth. If the steps were not taken, even the so-called exempt cooperatives would have to pay taxes. If the cooperative's net savings or net earnings were not distributed to the patrons, the cooperative corporation would be taxed like any other corporation. It is therefore a misnomer to refer to any cooperative as being exempt from federal income tax, although this is often done and this terminology is actually continued in the tax codes.

The basic assumptions upon which the single tax concept of deductibility by the cooperative and tax payment by the patron on patronage refunds began to break down in several court decisions in the 1950s. Patrons challenged the taxability of noncash allocations, and some of the courts agreed with them. In one case, the Court of Appeals for the 4th Circuit held that a patron did not receive income as a result of the patronage credit allotted, nor did the patron become entitled to receive anything that could properly be construed as income. To follow this line of reasoning, the result would be that someone was receiving income, but no one was paying tax on it. Critics of cooperatives and some cooperatives themselves advocated a tax through four sessions of Congress. Their positions and arguments differed, however, on how cooperatives should be taxed.

The critics argued that cooperatives should be taxed just like ordinary corporations without regard to patronage refunds. The representatives of cooperatives urged that the cooperatives be permitted to exclude their patronage distributions as in the past, with the patrons being taxed as had been intended in 1951. Congress essentially adopted the position of the cooperatives in the Revenue Act of 1962. There were, however, some very elaborate provisions added.

SUBCHAPTER T OF THE REVENUE ACT OF 1962

Subchapter T of the Internal Revenue Service Code provides the rules for the taxation of most cooperatives. Mutual savings banks and insurance companies and rural electric and telephone cooperatives are excluded.

In general, it preserves the single tax treatment that follows the cooperative method of doing business. This is done by providing the cooperatives under Section 1382(b) the right to reduce taxable income by an amount equal to the patronage dividends paid to members and patrons on the basis of the amount of business they did with the cooperative.

Very specific requirements, however, are set forth which the cooperative must observe to permit the deduction. The refund must be in the form of cash, other property, or qualified written notices of allocation. Cash and other property are easily understood, but the term qualified written notices of allocation often causes problems. As stated previously, a written notice of allocation is defined in Section 1388(b) as "any capital stock, revolving fund certificate, retain certificate, certificate of indebtedness, letter of advice, or other written notice, which discloses to the recipient the stated dollar amount allocated by the organization and the portion, if any, which constitutes a patronage dividend."

In order to "qualify" these written notices of allocation, several additional technical requirements had to be met. One of the most important is that the patron must consent to take the full stated amount of the notice into income in the year in which it is received. Section 1388(c)(2) provides that the required consent can be obtained in one of three different ways:

1. The patron may consent in writing. Such consent covers all patronage of the patron during the year as well as future years and until it has been revoked in writing by the patron.

2. The patron may consent by becoming a member of a cooperative association which has so-called consent bylaws. By becoming a member, the patron agrees to abide by the bylaws of the association, which in this case would normally require including the patronage dividends as ordinary income. However, it is extremely important to note that bylaw consent is effective only after the new member has obtained a copy of the bylaw together with a statement explaining its significance to him. There have been a number of instances recently in which revenue agents have disallowed deductions of patronage dividends on the ground that the new member had not actually received copies of the bylaws. In addition, it is important to remember that patronage of a new patron occurring prior to the date of his becoming a member (even if this patronage was during the same fiscal year of the cooperative) would not be covered by the consent bylaw.

3. *The third means available is through the so-called qualified check.* Such a check is defined in Section 1388(c)(4) as a check that is paid as part of a patronage dividend and on which there is clearly imprinted a statement that the endorsement and cashing of the check constitutes the consent of the payee to include in his gross income the stated dollar amount of the written notice of allocation which is a part of the patronage dividend or payment. Section 1388(d) provides that qualified check loses its status as a consent if it has not been cashed on or before the ninetieth day after the close of payment period for the taxable year of the cooperative. Because of this feature, use of a qualified check alone can be somewhat risky for the cooperative. However, if the check has been cashed, it qualified the entire amount of the patronage allocation. Its use is particularly helpful in situations where a cooperative makes patronage allocations to nonmember patrons and, of course, does not have the availability of the bylaw consent for such nonmember patrons.

One further and very important feature added by the Revenue Act of 1962 to the technical requirements for deductibility of exclusion of patronage dividends by a cooperative involves the minimum 20% cash payment. Under Section 1388(c), a patronage allocation is not qualified unless at least 20% of the patronage dividend of which it is a part is paid in cash. The legislative history suggests that this requirement was added to provide the farmer with the cash to pay the tax that would be due on the total amount of the patronage allocation. However, the 20% cash requirement exists for consumer items as well.

The regulations under Subchapter T also require that the notice to be qualified must state separately the percentage amount of the allocation attributable to patronage and that attributable to nonpatronage sources.

Lastly, with respect to qualification of patronage dividends, such allocations must be made within the "payment period," defined in the Code and Regulations to be within $8\frac{1}{2}$ months following the close of the fiscal year of the cooperative. The importance of this particular provision has been emphasized in a tax court decision wherein it was determined that no patronage dividend deduction was allowable due to the failure of the cooperative to pay such dividend within the $8\frac{1}{2}$ months. In one of the years in question in this case, the patronage dividend was paid 3 days after the required payment period. The taxpayer attempted to argue that a patronage dividend had been paid by virtue of the issuance of financial statements to the patrons which could have been used, together with the patrons' purchase receipts, to determine the allocable portion of the patronage dividend attributable to the patron. The court rejected this and all other arguments made by the cooperative, holding that the precise language of the Internal Revenue Code was controlling and the failure of the cooperative to issue a written notice stating the precise amount allocable to each patron caused the cooperative to lose its patronage dividend deduction.

PER UNIT RETAINS EXCLUSION

The Revenue Acts of 1966 and 1969 amended Subchapter T to permit deductions for qualified per unit retains. A per unit retain is defined as an allocation by the cooperative to a patron with respect to products marketed, the amount of which is fixed without reference to the net earnings of the cooperative. The per unit retain is deductible to the cooperative if the patron has either agreed in writing or is a member of a cooperative having a bylaw providing for such member–patron to include the stated amount in ordinary income. Unlike the patronage dividend, no part of the per unit retain need be paid in cash.

In summary, Subchapter T permits a cooperative to deduct patronage dividends and per unit retains if it complies with numerous technical requirements. Although the right to such exclusions has been recognized historically and conceptually as a "price adjustment," the IRS and the courts in recent years have tended to overlook this conceptual basis. Instead, they look to the provisions in the Internal Revenue Code as providing special statutory language setting forth precise guidelines which should be construed narrowly. This attitude will probably be with us for some time. This means that cooperative managers must be constantly alert and comply to the letter with the technical requirements of Subchapter T to assure deductibility.

PATRON TAX TREATMENT

In addition to the provisions relating to the tax impact on the cooperative, Subchapter T also sets forth in Section 1385 the tax impact on a patron receiving a patronage dividend or per unit retain. As would be expected, the patron recipient of cash and a qualified written notice of allocation (unless it is for a consumer item) must include the stated amount of such allocation in ordinary income for the year in which the patronage dividend was received.

The tax consequences to the patron of a nonqualified written notice of allocation, however, are a bit more involved. For example, issuance of a written notice within the $8\frac{1}{2}$ month payment period not consented to by the patron or not including 20% in cash results in the notice being nonqualified. Also, use by a cooperative of insufficient language on a purported qualified check could result in the allocation being nonqualified.

SECTION 521

Thus far, in the discussion of Subchapter T, no specific reference has been made to agricultural cooperatives. The reason is that Subchapter T applies (with a few exceptions noted earlier) to all corporations operating on a cooperative basis. This would include consumer cooperatives, retail gasoline stations buying together, railroads jointly operating a bridge, and many other examples of nonfarm cooperatives. Of course, it also covers the taxation of agricultural cooperatives.

Agricultural cooperatives, unlike other cooperatives, also have the option of operating in such manner as to comply with another section of the Internal Revenue Code, Section 521. If they do, they are permitted under Subchapter T [Section 1382(c)] to deduct, in addition to the allocations of patronage earnings, the following two items:

1. Any amounts paid during the taxable year as dividends on its capital stock.
2. Earnings from nonpatronage business, for example, rental income and earnings on business done with or for the U.S. Government, if such patronage income has been allocated to members and patrons on the basis of their patronage.

The tax saving to the cooperative having either substantial amounts of nonpatronage earnings or paying dividends on capital stock can obviously be very significant.

To be eligible for these special tax benefits under Section 521, a cooperative must meet the following requirements:

1. Be a farmer, fruit grower, or like association.
2. Be organized and operated on a cooperative basis either for (a) the purpose of marketing the products of members or other producers

and turning back to them the proceeds of sales less the necessary marketing expenses on the basis of either the quantity or the value of the products furnished, or (b) the purpose of purchasing supplies and equipment for the use of members or other persons and turning over such supplies and equipment to them at actual cost, plus necessary expenses.

3. Organizations having capital stock must (a) limit the dividend rate on such stock to the greater of 8% per annum or the legal rate of interest in the state of incorporation on the value of the consideration for which the stock was issued, and (b) substantially all of such stock (other than nonvoting preferred stock) must be owned by producers who market their products or purchase their supplies and equipment through the association.

4. Transactions with nonmembers must not exceed the value of transactions with members.

5. In the case of a purchasing cooperative, transactions with persons who are neither members nor producers must not exceed 15% of the value of all transactions.

In recent years, there has been a considerable amount of activity in connection with the interpretation and construction of the various eligibility requirements. Several rulings have been made, but a few of the key issues are as follows:

1. Whether the association constitutes a "farmer, fruit grower, or like association." The IRS has held that the term "like associations" is limited by the earlier references to farmers and fruit growers and includes only associations of farmers or others engaged in occupations that could be considered farming.

2. Marketing and purchasing. An extremely important consideration in eligibility is the construction of what constitutes marketing or purchasing on behalf of members. This is particularly significant in the marketing area where farmers may operate on an agency basis and occasionally form cooperatives for purposes related to, but not necessarily constituting, a marketing function. A cooperative that leases its facilities, for example, is said to have leased to perform marketing functions.

3. Substantially all stock (other than preferred stock) held by producers who market their products or purchase supplies through the association.

WHO IS A PRODUCER AND
WHAT IS MEANT BY SUBSTANTIALLY ALL?

An important consideration running through Section 521 involves a definition of producer. Inasmuch as Section 521 applies to farmers' associations and also requires that substantially all stock be held by producers, an acceptable definition is important. A major revenue ruling on this question provides in general that a person is a producer if the owner or tenant bears the risk of production and cultivates, operates, or manages the farm for gain or profit. A person receiving a rental based on farm production (either in cash or in kind) is also considered a producer. The landlord is not considered a producer if a fixed rental is received. A stockholder in farm cooperative corporations is not considered a producer by virtue of stock ownership.

Perhaps the most intensive effort on the part of the IRS in the area of interpreting Section 521 has come in connection with the question as to whether substantially all of the voting stock of a cooperative is held by producers who market or purchase through the association.

The law does not define specifically what percentage of the shares must be held by producers. A rule of reason was said to prevail, however, and some guidelines have been established in the cases. The IRS has maintained administratively for some time and has now published a revenue ruling that substantially all means at least 85%.

CURRENT PATRONS

In addition to the 85% requirement, the IRS now requires that these producers must also be active producers currently patronizing the association. In one ruling, it was concluded that on a current basis, as used in Section 521, means actual yearly participation. Continuation as a producer without actual yearly participation is not sufficient. In other words, even if 85% of the capital stock is held by active farmers, these shareholders must patronize the cooperative each year to protect the Section 521 exemption.

This means that the cooperative must review its membership list annually. This will be a particularly onerous burden to Section 521 cooperatives who have retired shareholders, considerable turnover in members, or members who do not patronize the cooperative each and every year.

Another major problem in connection with maintaining Section 521 status involves the IRS prohibition against marketing, except in unusual circumstances, agricultural products for or on behalf of nonproducers. Occasionally cooperatives have sought to reduce the cost of doing business for members by processing commodities obtained from nonproducers. Also, some cooperatives such as an operating dairy cooperative having fluid milk distribution to retail customers occasionally find that the consumer desires some nonfarm products be delivered along with the fluid milk. The position of the IRS is that such marketing can be justified only if it is clearly incidental to the overall marketing of producer products and in no event should exceed 5% of the total sales. In the case of wholesale sales by a marketing cooperative, the IRS has refused to permit even 5% of total sales.

EQUALITY OF TREATMENT
BETWEEN MEMBERS AND NONMEMBERS

One of the most significant differences in the operation of a cooperative qualifying under Section 521 as opposed to a so-called nonexempt cooperative is that the Section 521 cooperative must allocate patronage earnings to members and nonmembers alike. A nonexempt cooperative need not pay patronage dividends to nonmembers.

TAX EXAMPLE—
SECTION 521 COOPERATIVE

After studying the summary and definitional portion of the chapter and the more detailed elaboration that followed, let's now examine a hypothetical example of a cooperative's income statement and how it would be handled from the standpoint of federal income taxes. An income statement is shown in Table 13.1.

We have assumed in this case that this cooperative has met all the Section 521 requirements. These can be reviewed in both the summary area and the detailed elaboration. If these requirements are met, the cooperative can pay federal income tax on the net margin less any patronage dividends paid, qualified per unit retains, and dividends paid on capital stock. The cooperative chose to pay income taxes on the nonpatronage business in this case, although it could have been allocated to members and patrons on the basis of their patronage with the cooperative. This means that this cooperative would pay federal income tax on the $2000 shown as residual and which came from nonmember–

TABLE 13.1
Hypothetical Cooperative Income Statement and How Handled
for Federal Income Tax Purposes[a]

Sales		$200,000
Cost of goods sold		120,000
Gross margin		$ 80,000
Costs		60,000
Net operating margin		$ 20,000
Income—Other sources (nonpatronage margin)		10,000
Net margin		$ 30,000
How net margin was allocated		
Patronage dividends		$ 8,000
Cash	$2,000	
Noncash	$6,000	
Per unit retains		$ 4,000
Stock dividends		$ 6,000
Other income (nonpatronage margin)		$ 10,000
Residual (nonmember–nonproducer)		$ 2,000

[a]See text discussion for details in regard to how the statement
was handled for federal income tax purposes.

nonproducer business. Such business, it is recalled, might come from
interest, rental income, or business done with the government.

TAXING INCOME TO THE PATRON

A patron of the cooperative receiving such allocations, as indicated,
must include in gross income the stated dollar amount of the written
notice of allocation which is a part of the patronage dividend or payment. This would be in the year received by the patron if it is qualified.
If it is not qualified, the cash portion would be included in the year
received and the noncash portion in the year in which it is redeemed.

In the case of per unit retains, they would be included in the year
received if qualified and in the year redeemed for cash if not qualified.
Stock dividends would be included as income in the year received and
earnings from nonpatronage business would be handled the same as
patronage dividends when they are allocated to patrons.

In essence, what has been demonstrated here is in keeping with the
single tax concept. The cooperative has distributed its net margins or
savings to its patrons in proportion to the amount of business they have

done with the cooperative—their patronage. It did this because of an obligation to do so which existed at the time the trade or patronage occurred. It was thus a preexisting obligation, and in doing this it was legally entitled to exclude those net savings from its income. Actually, the cooperative had had no income, except in this hypothetical case the earnings from nonpatronage business, since the amounts involved belonged to the patrons from the beginning. However, they became income to the patron because the cooperative had distributed or allocated them and were taxed at that point.

TAXING INCOME TO
PRIVATE CORPORATIONS AND STOCKHOLDERS

In contrast to this procedure, and this is where arguments arise, the private corporations would pay income taxes on the net margin, in this case, $30,000. Stockholders in the corporation would declare any dividends that were declared by the corporate board and paid to them in accordance with the number of shares of stock of the corporation they held. Thus, the double taxation charge arises.

AGRICULTURAL COOPERATIVES—
A SPECIAL CLASS

It is remembered that Subchapter T applies, with a few exceptions, noted earlier in this chapter, to all corporations operating on a cooperative basis. It includes agricultural cooperatives, but it also covers the taxation of nonagricultural cooperatives. Only agriculatural cooperatives, however, have the option of operating in such a manner as to comply with another section of the Internal Revenue Code, Section 521. If they do comply with these specific and strict requirements, they are permitted to deduct the patronage earnings which have been allocated. They may also deduct dividends paid on their capital stock and earnings from nonpatronage business such as rental income and earnings on business done with or for the U.S. Government if such income has been allocated to members and patrons on the basis of their patronage.

THE CHALLENGE

As indicated earlier, our objectives in this chapter were aimed at providing sufficient information in regard to taxing of agricultural cooperatives for the student to begin building an understanding of some of the issues involved. Much more study would be necessary to become qualified as an expert in corporate taxation, including private corporations, all Subchapter T cooperative corporations, and the special case of agricultural cooperative corporations which have an option to comply with specific requirements under Section 521 of the Internal Revenue Service Code. If they do comply, they are then permitted to operate under the single business tax concept in its fullest meaning.

As has been stated explicitly at several points and as has been implicit at all times, it is hoped that an analytical, clinical, and questioning posture would be assumed at all times. It is hoped that no position would ever be taken in regard to any issue without insisting that as many relevant facts as possible are brought to bear on the issue—that positions based upon emotions and preconceived notions would be eschewed as being unprofessional. It is hoped that ideas and concepts would be subjected to constructive critical evaluation. It is hoped that such a posture becomes second nature and is automatically assumed whenever any issue, cooperative corporation taxation policy or whatever, is being considered. This is the essence of professionalism. No other posture can serve this end.

Many areas of relevance have been mentioned in this book which may be revisited in seeking to establish a position we are willing to defend regarding the taxation issue—industry structure, economic power of individual firms as related to structure, the ownership, control, and reason for being of various kinds of business organizations, to name a few. The kind of product produced by agriculture and its essential nature are also relevant.

At the same time, we must remember that a sound, fair, and strong federal tax system is essential to our future. Without such a system, we cannot render the public services necessary to enrich the lives of our people and further the growth of our economy.

Such a system must be adequate to meet our public needs. It must meet them fairly by calling on each of us to contribute our proper share to the cost of government.

It is within this overall context that we bring relevant information to bear upon this issue in a professional manner and seek a defensible position.

REFERENCES

Baarda, J. R. 1982. State Incorporation Statutes for Farmer Cooperatives. Cooperative Information Report No. *30*. Farmer Cooperative Service, United States Department of Agriculture.

Morris, R. K. 1984. Cooperatives, The Law and Taxes. NICE, Bozeman, MT.

U.S. Department of Agriculture, FCS. 1976. Federal Income Taxes, Legal Phases of Farmer Cooperatives, Part 2. FCS Information *100*.

U.S. Department of Agriculture, FCS. 1981. Cooperative Financing and Taxation. Cooperative Information Report *1*, Section 9.

U.S. Department of Agriculture, FCS. 1983. Antitrust Laws, Legal Phases of Farmer Cooperatives, Part 3. FCS Information *100*.

TO HELP IN LEARNING

1. Quiz 10 of your peers. Use this question, "What, if any, taxes do agricultural cooperatives pay?"

2. Conduct a bag lunch seminar on the topic: "Taxing of Cooperative Corporations and Private Corporations—The Rationale Involved."

3. Nonprofit means no net income, it would seem. What is the controversy about?

4. Talk with a federal income tax expert and determine what position is taken on taxing agricultural cooperatives and the reason(s) behind it.

5. Talk with the general manager of a cooperative corporation. What views on taxing cooperatives are held and why?

6. Talk with an executive in a private corporation. Determine what views on taxing cooperatives are held and why?

DIRECT QUESTIONS

1. Do cooperatives pay taxes? Discuss.

2. What is the "single tax" concept?

3. What is the rationale for the single tax concept?

4. What does "qualified for exclusion from income tax obligation of the cooperative" mean?

5. How can patronage dividends be qualified?

6. What is a qualified written notice of allocation? What form can it take?

7. How can per unit retains be qualified for exclusion from income for taxing purposes?

8. What are Subchapter T and Section 521 of the Internal Revenue Code?

9. How can a cooperative qualify for exemption from taxation under Section 521?

10. How does a cooperative handle nonpatronage business income such as interest under Section 521?

11. If an allocation is unqualified, what happens?

12. How are written notices qualified?

13. What is the rationale behind the requirement that 20% of the patronage dividend must be paid in cash?

14. Does the 20% cash provision apply to per unit retains allocated to patrons?

15. When a patron receives a qualified refund, what is the income tax obligation?

16. If an allocation is unqualified, what income tax obligation is there?

17. What are the additional two items which Section 5?1 cooperatives are permitted to exclude from income for tax purposes?

18. What is meant by "substantially all" stock must be held by active producers?

19. Receiving a fixed rental by a landlord is qualification for being considered a producer for Section 521 purposes. Yes or no? Why?

20. Who are "current" patrons of a cooperative?

21. What is an "exempt" cooperative?

22. What is meant by a "preexisting" obligation on the part of a cooperative and what is its significance in taxing?

23. Does Subchapter T apply only to agricultural cooperatives?

24. Does Section 521 of the IRS Code apply only to agricultural cooperatives?

25. Private corporations are taxed differently than cooperative corporations in what way?

TYING-TOGETHER QUESTIONS

1. There are those who argue that cooperatives should be taxed just like ordinary corporations without regard to patronage refunds. Prepare a short position paper which reflects and supports your views on this.

2. Most agree that a sound and strong federal income tax system is essential. It must be adequate and it must be fair in that it calls on each of us to contribute our proper share to the cost of government. Draw upon every facet of the analytical framework which we've tried to develop and plot the case for procedures used in taxing agricultural cooperatives.

3. React to this statement, "Private corporations should be taxed in the same way as cooperative corporations."

4. Prepare a rationale for taxing cooperatives based upon the principles involved.

5. Prepare a rationale for taxing cooperatives and private corporations based upon their differences.

6. Which approach appears to be more soundly based? Why?

14

Cooperatives and Federal Commodity Marketing Orders

Perhaps the study of no other institutional arrangement addresses questions and issues relating to the structure of agriculture vis-à-vis other sectors of our economy better than does the evolution of our federal commodity marketing order system. Competition, the demand characteristics of food products, the biological nature of agricultural production, and the public interest are all involved.

Examination of the underlying economic rationale for considering and ultimately adopting such a mechanism is an interesting and fruitful exercise which can establish more firmly in our minds some of the concepts covered in Part I of this book relating to these areas. Should there be further need for help in reaching that comfortable position regarding the why of agricultural cooperatives which we've established as one of our goals, studious and contemplative examination of the material in this chapter should prove to be useful.

Federal marketing orders are so cooperative in nature and so closely tied in with cooperative action that they may be considered in almost the same breath. Their underpinnings, backgrounds and reason for being, basic methods of operation, and objectives are so similar that they could almost be handled under the same legislative provisions. However, they are legislatively different. Their background and legislative provisions along with their use by cooperatives will be covered in this chapter. The high degree of complementarity between federal orders and agricultural cooperatives and also the supplementary nature of orders in helping cooperatives achieve their goals will provide the focus.

The thrust of the effort here will be aimed at exploring the why of federal marketing orders, with some concern for their makeup. The how portion will be explored in Assignment 12 of your Cases in Cooperative Marketing.

THE ROOTS OF THE PRESENT SYSTEM
OF FEDERAL MARKETING ORDERS

The roots of the present system of federal marketing orders go far back to the evolution of our free market institutional arrangements and our traditional concepts of competition and governmental regulations. Our leanings were in the direction of competition as it was defined in a previous chapter and toward free markets. Price and competition would provide whatever regulatory force that might be needed if permitted to do so by our following a hands-off policy of "laissez-faire."

Many factors entered the picture, however, and began the process of erosion of the system that reflected our traditional concepts of what was economically desirable. Increasing urbanization of our country and advancing technologies almost dictated large-sized firms as well as the growing concentration of many local small handlers of agricultural commodities into industries made up of a few large and economically powerful firms.

All three of these factors have elements that go far beyond what farmers or the U.S. Department of Agriculture could do from the standpoint of effective policy. Each was important in contributing to the institutional arrangements that finally evolved. However, the third factor, the trend toward the concentration of the various businesses that sold inputs to agriculture and bought its output into the hands of large-scale organizations, will provide focus for our examination of federal marketing orders.

CONDITIONS FACED BY AGRICULTURE

Most of agriculture shared the comparative prosperity of the World War I period. But when, after the market crash of October 1929, the economy moved into a period of general recession or depression, farmers were convinced that their local efforts needed to be complemented by some form of government reinforcement. This created a crisis in agriculture, especially in areas such as fruits and vegetables where producers had been induced by the high wartime prices to make new plantings and which were now coming into production. Prices dropped, and since resources had been committed to production of crops, the situation was made worse by the added production. The boom and bust scenario characterizing conditions at that time was being played out in devastating form for agriculture in general.

WHAT DID AGRICULTURE TRY TO DO?

As has been shown before, the history of the agricultural cooperative movement in the United States shows clearly that these associations arose as an attempt to do something about the income situation of their members. These early cooperatives pioneered several procedures for coping with the problems they faced. In the hard school of practical experience they had picked up considerable understanding of the mechanics of the marketplace and the economic principles involved. They were well aware that if supply outruns demand at a prevailing level of prices, competition among handlers engenders price cutting, and the economic weaknesses of individual farmers force them to accept whatever is offered. Similarly, when supplies are short, producer groups will try to force prices up as high as the traffic will bear. Buyers of the commodities will then scout around for cheaper products. The end result of this cold war setting was frequent demoralization of the business of providing an adequate supply of the product involved to the ever-growing metropolitan markets. A stable peace or at least peaceful coexistence was obviously needed, but there seemingly was no way of bringing this about. In the case of the dairy sector of the agricultural industry, such a situation was sought under the slogan "orderly marketing."

FRUIT AND VEGETABLE GROWERS

In the case of fruit and vegetable producers, attention turned to the development of programs to regulate the quantity and quality of fruits and vegetables marketed. A few of the stronger cooperatives attempted unilateral regulation of the marketing and growing of fruits, but achieved little lasting success and all eventually failed. A common limitation leading to their failure was their inability to induce or maintain participation by a sufficient number of producers and handlers of the products. Since nonparticipants enjoyed the same benefits as participants while avoiding the costs involved, there was little incentive to participate. This situation, which is tied directly to the structural arrangement in agriculture, is referred to as the "free rider" problem, and it characterizes most voluntary and self-help programs in agriculture.

THE DAIRY SUBSECTOR

Dairy farmers continued their efforts to help themselves and in the process became more and more knowledgeable about market conditions and concepts and what was involved. They sought to improve their bargaining power by amassing a large supply of their product and hiring skilled merchandisers to sell it. They devised concepts and practices of "classified" pricing to the dealer according to the use to which the milk was put and of "pooling" returns to members so that all would share equitably in both the higher returns from fluid milk and the lower returns from manufactured products or surplus uses. However, they could not fully enforce and implement the classified pricing system they devised because they could not prove how the dealers used their milk in the absence of an effective audit of handler records showing the amount of milk in the absence of an effective audit of handler records showing the amount of milk received, what they paid for it, and what they did with it. Furthermore, some producers would not join the cooperative and thus would not participate in the pooling and classified pricing plans devised by the producers in an attempt to overcome some of the problems involved. The spectre of the free rider arises again to help frustrate the efforts of some.

Despite these very conscientious efforts on the part of farmers to help themselves through their cooperatives and bring greater stability to conditions, they were frustrated when they failed. Economic power related to structure on the side of buyers, and the free rider problem with their own group made the task impossible for the self-help routes. They became more and more convinced that their local efforts needed to be complemented by some form of government reinforcement. This would almost inevitably suggest some sort of involvement of the government, but before we move directly into what this involvement was, let's review some of the concepts to which we have been exposed earlier. Such a review may be helpful in thinking through the position we are trying to establish in regard to whether cooperatives are justified and perhaps to either solidify it or raise further questions about it.

SOME RELEVANT CONCEPTS

Several concepts are of relevance in our study of the circumstances that led to enactment of legislation which provided for federal marketing orders. These include the concepts of (1) administered prices, (2) orderly marketing, (3) agricultural adjustment, and (4) the public interest.

An understanding of these concepts as they tie in with economic conditions and the legislation that emerged can be helpful, not only in why the legislation was enacted, but how it works.

ADMINISTERED PRICES

The first concept which is relevant for our overall purpose in this course and for the immediate purpose in this chapter is that referred to as administered prices. The domain of this species of prices spreads over the area between the theoretically automatic prices of very small-scale or atomistic competition in a predominantly free market with many buyers and sellers and the completely controlled prices of a monopolistic setup. As was discussed in Chapters 1 and 2, pricing behavior of a type covered by the term administered prices is simply descriptive of the way business is done in highly centralized industries such as steel and automobiles.

The distinctive feature of administered prices is that the price maker has a significant ability to determine or influence the flow of supplies, products, and related services through the adoption of a price policy. Goals of unit price and/or overall profits are set up and then supply is adjusted toward the attainment of that price–profit objective.

The degree of supply control on the part of the individual agricultural producer is nil as compared with that of the suppliers of inputs and buyers of outputs. Cooperatives, or groups of producers, had attempted to engage in a limited form of administered pricing when, for example, they attempted to implement a policy of allocating products between different uses, classified pricing, but they failed in the attempt. They had also developed other ingenious methods, such as pooling and seasonal pricing devices, to deal with the chaotic conditions that prevailed during and after the depressed conditions of the 1930s, but their efforts to implement them also failed. Was the stage now set for asking government to play a role in making it possible for agricultural producers to administer prices through some mechanism?

ORDERLY MARKETING

The next concept relevant for our purposes is that of orderly marketing. This concept should be considered while keeping in mind what is involved in administered pricing, just discussed. It should be reviewed again once the concept of the public interest has been discussed. What is orderly marketing and what is its relevance for our purpose?

The classical doctrine that the unregulated competition of the free market would act as an automatic adjuster of both price and production had some merit in its day of small-scale business operators. But as investments and fixed costs necessary for the adoption of new technology have grown, such a doctrine has become close to meaningless.

The meaning of the concept of orderliness in the economic sense is still not fully agreed upon by everyone. It seems reasonable, however, that if consumers of agricultural products are to have an orderly supply, there must be orderly production. Further, if there is to be orderly production, both efficient and remunerative, there should be orderly provision for assembly and distribution of the products and for dependable and equitable contract relations between producers and buyers of their products.

Early cooperative efforts were directed at the achievement of some degree of orderliness in marketing. They sought to get away from severe and often unpredictable swings from surplus to shortage of the product, often within the year, and to secure a permanent and dependable membership who would loyally support the policies and programs aimed toward the benefit of the whole group. In addition, they sought to build up a vested interest in a desirable market situation in which the interests of all could be served and to protect this interest from intrusion by others. Finally, they sought to hold prices fairly steady for considerable periods of time and to respond to changed economic conditions such as the general price level, farm costs, pressures from alternative supplies, or the like, rationally and gradually.

They tried to do all of these things, but they failed. They then started looking to government as an accomplice in their efforts. Were they justified in doing this and what has emerged?

AGRICULTURAL ADJUSTMENT

The third concept which will be helpful to us in understanding what the economic circumstances were, what part they played in bringing about legislation, what the legislation was designed to do, and perhaps in understanding how it works, is that of agricultural adjustment. In a broader sense, this could be viewed as economic adjustment, but for our purposes we are concerned with agriculture as it attempts to adjust to the economic circumstances that surround it.

Had we pursued our legislation history or background a bit further, we would remember that the original statute from which the federal order system sprang was referred to as agricultural adjustment. The first legislation of this nature was enacted soon after the Great Depression of 1929 and was called the AAA, the Agricultural Adjustment Act. This Act set up an apparatus for improving the lot of the farmers by helping them into better equilibrium over time with market demands that were relatively inelastic. The broad goal of improving farm incomes was centered on the word parity, and the concept of parity was incorporated into federal order legislation and is still in use today.

In the unique system of free enterprise production and orderly marketing of agricultural products which was inaugurated and which we are still trying to perfect, the position sought lies somewhere between rigorous government control and reckless mayhem among partisan groups, groups often of considerable differences in bargaining power and circumstances. Eventually, the Secretary of Agriculture was to be placed into the role of the moderator of a process that supposedly was dedicated to promotion of the public interest. This process was designed to bring about a rational adjustment of several parts of agriculture, milk and several other commodities, and was referred to by John D. Black as "assisted laissez faire." This step, along with the Capper–Volstead Act which came a bit earlier, was designed to tailor the master concept of individual competitive enterprise to the diverse and changing conditions of large-scale operations and advancing technology.

It is recalled from our discussions in Part I of this book that the very essence of what we have learned about the nature of economic equilibrium in the process of growth and also of stability is summed up in the word competition. Freedom of movement and of entry and exit provide an opportunity for producers to embrace the most profitable economic opportunities they can discover or create. Consumers, in this process, will have access to the best and most economical sources of want satisfaction, and distributors will have incentives to find the most economical sources of supply and will create the most efficient means of reaching them and serving users. The nation's resources will be allocated to the most economic uses.

This basic notion and principle of competition as an organizing force remains, but the manner of its application has changed over time. In today's situation in which highly capitalized and big unit organizations have become the norm, we have graduated from atomistic competition to monopolistic competition. It is characterized by a rather high degree of control by management. A given supplier has sufficient control over volume in the market that it can exert significant influence on market price. This is not monopoly, but neither is it blind competition among powerless individuals.

It is a matter of public and also of private concern that this process of supply control and pricing be conducted in such a way that the essentials of the free enterprise system with regard to optimal allocation of productive resources are maintained. Administered pricing should be the ultimate expression of the idea of orderly marketing, properly guided by the trained intelligence of production and marketing experts. The general public interest should be promoted.

As a part of that interest, it must be recognized that producers of agricultural products have claims. They have immobile investments which, of course, in some cases may have been ill-conceived. Adjustment or reform must not move so fast toward some ideal that the process itself becomes a factor of economic chaos and thus a part of the problem.

It is with this economic background review that the reader is encouraged to consider the marketing agreement legislation under which federal marketing orders are promulgated and operated. Why did we graduate from the concept of atomistic competition to one called monopolistic competition and were we justified in doing so? Is the public or general interest being served by that movement and how might it have been served had we not moved? Let us now move to a more definitive consideration of what is involved in the concept of the public or general interest.

THE PUBLIC INTEREST

The traditional concept of the public interest stems from the objective stated in our Constitution to promote the general welfare. Use of this term in the Agricultural Adjustment Act of 1933 and later in the Agricultural Marketing Agreement Act of 1937 under which federal marketing orders operate was clearly intended to allay any fears in the minds of Congress, editors, or the general public that class legislation was being enacted which could be used to the detriment of other segments of society.

Congress and the general public had come to recognize on a wide scale that farmers had been peculiarly disadvantaged after the war by price declines which were more severe and more persistent in the case of farm products than for industrial products. The AAA of 1933 recognized this when it stated that such declines impair the purchasing power of farmers, destroy the value of agricultural assets, and affect the national public interest. By implication, at least, the Congress was saying that the national public interest could not be fully secured if the interests of agriculture were neglected.

The interest of the second component of the concept was not overlooked. It was declared to be the policy of the Congress to protect the interests of the consumer by gradual corrections of deteriorated price levels to agricultural producers, and no action should be taken which had for its purpose the maintenance of prices above the level established as being appropriate (i.e., parity). It is recalled that Adam Smith pointed out that consumption is the sole end and purpose of all production. It is also recalled that Section 2 of the Capper–Volstead Act was prepared with consumer protection in mind.

Producers, of course, want as high a price as they can get and consumers want as low a price as possible. Distributors and processors want their margins to be as high as possible. However, it is contrary to the public interest to have prices so high that the consumer is exploited and the use of essential food elements is restricted. It is equally against the public interest to have agricultural prices so depressed as to impair the operating efficiency of the producer and destroy the value of the assets involved. It is also in the interest of the public that the margins of those between producers and consumers be adequate for providing facilities in keeping with technological progress and for giving the best service to all concerned.

All of this suggests strongly that the interests of producers, handlers, and consumers as parts of the general public interest should not stand in a confrontational mode with respect to each other. Their interests are, in fact, mutual. It is not in the long-run best interests of agricultural producers to price their products so high that consumer needs are not being met. By the same token, it is not in the long-run best interest of consumers to have food product prices so low that the assets of the producer are jeopardized and, if carried too far, could result in food products not being available to them or available only at unduly high prices. A prime criterion of any legislation that establishes policy and the program designed to implement the policy is that they should be in the public interest. The producer, handler, and consumer are all parts of the public interest that must be served.

It is within this environment that the concern with instability, as reflected in disastrously low farm prices, resulted in considering legislation by the Congress which was aimed at bringing about more orderly marketing and greater economic stability. There was general agreement that restoring and maintaining those conditions would not only be in the best interest of agricultural producers, but would also be in the best interest of all—the public interest. The student is again strongly encouraged to keep these concepts and their implications in mind as we delve further into what federal marketing orders are and how they work. Did the circumstances justify the enactment of this type of legislation and is the public interest being served by it?

FEDERAL COMMODITY MARKETING ORDERS

Authority

Legislative authority for federal commodity marketing orders rests in the Agricultural Marketing Agreement Act of 1937 as amended. The law has been changed slightly and coverage has been extended to additional commodities, but its basic provisions have remained unchanged since it was enacted.

How Does a Federal Marketing Order
Come into Being?

Marketing orders are issued by the Secretary of Agriculture after a notice and opportunity for hearing on each proposed order. The request for the Secretary to consider the use of an order usually comes from a group of producers in the area, usually a cooperative, and cites problems in the area which are alleged to exist. A proposed marketing order is submitted along with the request for a hearing.

The Secretary of Agriculture reviews the request for a hearing to consider the possible use of an order. In most cases, a marketing specialist will be sent into the proposed marketing area to examine firsthand the bases for the statement that problems exist in the market and that the use of a marketing order is needed to overcome the problem. The marketing specialist will talk with cooperative leaders and other producers, processors of the commodity for which an order might be considered, and with consumers. Should the specialist concur with those who had petitioned the Secretary on the basis of the findings, a recommendation will be made to the Secretary that a hearing be held in the area for the purpose of considering the use of a federal marketing order in overcoming the problems.

The Secretary will then announce and have published in the Federal Register a notice that a public hearing will be held, beginning on a specific date at a specific hour at a specific place, for the purpose of hearing evidence of the existence of a problem(s) in the area and of the possible use of a federal marketing order in overcoming it. The public, producers, handlers, and consumers—everyone—is invited to participate in the hearings.

The Hearing Process

The hearing is convened at the announced time and place. It is presided over by an attorney, called the hearing officer, who is thoroughly familiar with the provisions of the Marketing Act of 1937 and the procedures to be followed in conducting the hearing.

Cooperative leaders are always present at such hearings, along with their attorneys. Handlers, along with their attorneys, are always present. Over the past few years, a few persons appear at the hearings as consumers. Anyone who has any interest in the proceedings and/or procedures is welcome to attend and to testify if they wish. All of us, of course, do have an interest.

The hearing officer, at the announced time, opens the proceedings with a statement regarding their purpose and the procedure to be followed. Testimony relating to whether the commodity in question is traded in or moves in interstate commerce is requested, and after the one who is to give the testimony is duly sworn, it is given. This is a requirement for the use of federal legislation, and in today's economy, most commodities qualify in this regard. In the case of a federal milk marketing order hearing, this testimony, along with other relevant data and information, is usually given by a representative of a cooperative or others who have evidence that such movements do take place. A court reporter is always present and a verbatim transcript of all testimony is made, with exhibits being received and duly marked to be included as a part of the official record of the hearing.

Once the legal basis of the hearing regarding the movement of the product in interstate commerce has been established, the hearing proceeds. Witnesses are presented by all parties, producers, processors, or others who have an interest. They are announced by the attorney for the party seeking to testify and the witnesses are then sworn to "tell the truth, the whole truth, and nothing but the truth" by the hearing officer. The witnesses then proceed with the testimony to be given. The testimony may be in support of or in opposition to any proposal(s) being considered. Once the witness has finished testifying, there may be direct examination by the attorney or others of the group that is represented for the purpose of clarifying any points that had been made. The witness is then subject to cross-examination by attorneys or others from opposing sides should they wish to question the testimony presented. All witnesses for that group are presented in this same manner, are heard directly, are cross-examined, and finish their presentation of evidence.

The other side or group, usually handlers or processors of the commodity in question, then begins the process of presenting evidence in an attempt to support their position regarding the question at hand. Again, each witness is sworn by the hearing officer, is allowed to present testimony, and is subjected to direct and cross-examination by the attorney and by others, including marketing specialists who are at the hearing representing the Secretary of Agriculture, should they have questions regarding the evidence or any other relevant area.

This procedure is followed in very exacting detail until each and every group, side, person, or otherwise has had an opportunity to be heard and to enter testimony into the record. After making a determination that everyone who wished to be heard has, in fact, been heard and that all testimony is in the record, the hearing officer declares the hearing closed. It is stated that the word-for-word record that has been made of the hearings, along with all the exhibits, properly numbered and identified in the record, are to be taken to Washington, D.C. for study and analysis by the U.S. Department of Agriculture marketing specialists. After such study and analysis has been made, they are then to make a recommendation to the Secretary of Agriculture in regard to whether a marketing order should be issued and the detailed provisions it should contain if one is issued. The hearing officer also states that those who wish to have a copy of the transcript of the hearings may purchase one at a stated price per page.

The Next Steps

After study and analysis of the hearing record, the marketing specialists make a recommendation to the Secretary of Agriculture. Should the recommendation be that an order should be established for the area, an order is formulated incorporating the suggestions made at the hearing, and a tentative decision regarding the order and the provisions is sent to the interested parties who attended the hearings and is published in the

Federal Register. A period of time, usually about 30 to 60 days, is given for filing with the Secretary, any exceptions anyone has regarding the order. The exceptions are duly examined by the marketing specialists, any revisions in the original order which they deem appropriate are made, and a final order is issued, published in the Federal Register and sent out for approval in a referendum by producers of the commodity. If two-thirds of the eligible producers of the commodity voting or producers with two-thirds of the volume of the product represented in the referendum vote in favor of the order, it goes into effect. This is in the case of all commodities except milk. In the case of milk, two-thirds of the producers of milk in the area must approve the order if it is to go into effect and if it provides for a marketwide pool. In the case of an individual handler pool, which is rather infrequent, three-fourths of the milk producers must approve the order. Bloc voting is used in that cooperatives vote their membership.

Terminating an Order

An order may be terminated or suspended by the Secretary of Agriculture if he finds it is not fulfilling the intent of the Agricultural Marketing Agreement Act of 1937 or that termination is favored by a majority of the producers of the commodity as specified in the Act.

Who Is under an Order?

Once issued, a marketing order is binding on all handlers of the product in the marketing area covered. This eliminates the problems encountered with marketing agreements and other similar arrangements where participation is voluntary. Such programs failed because of the free rider problem, the ability of nonparticipants to benefit without meeting the requirements and sharing the costs of the program.

How Many Federal Marketing Orders Are There?

In mid-1981, there were 48 federal marketing orders for fruits, vegetables, and specialty crops under the provisions of the Agricultural Marketing Agreement Act of 1937. The peanut program involves a marketing agreement, but no marketing order. More than half of the fruits and tree nuts produced in the country, measured in value terms, and about 15% of the vegetables are covered by these programs. These covered commodities had an estimated farm value of $5.2 billion in 1980. This represents about 8% of total farm receipts from crop sales and about $23 per person in the United States. A breakdown by type of fruit and vegetable orders by decade is shown in Table 14.1.

TABLE 14.1

Numbers of Federal Marketing Order and Agreement Programs in Effect on January 1, by Decade, 1940–1980 and 1981

Type of commodity	1940	1950	1960	1970	1980	1981
Citrus	2	4	6	9	9	9
Other fruits	5	8	10	16	16	17
Dried fruits	0	2	3	3	3	3
Tree nuts	1	3	3	3	3	3
Potatoes	1	9	8	7	6	6
Other vegetables	7	2	6	6	7	7
Peanuts, hops, and spearmint oil	1	1	0	2	2	3
Totals	17	29	36	46	46	48

Sources: National Commission on Food Marketing, Federal and State Enabling Legislation for Fruit and Vegetable Marketing Orders, Evolution and Current Status, Supp. 3 to Tech. Study No. 4, June 1966; Foytik, Jerry, "Marketing Agreements: Fruits and Vegetables" in Benedict, Murray R. and Oscar Stine, The Agricultural Commodity Programs: Two Decades of Experience, The Twentieth Century Fund, New York, 1956; and USDA, Agricultural Marketing Service records.

In the case of milk, there were 44 federal marketing orders in 1985. Federal order receipts of milk represented about 70% of total milk marketings, and 81% of all the fluid milk produced in the United States was handled under the provisions of federal marketing orders in 1982. It is safe to say that all the milk produced in the United States is either directly or indirectly affected by these orders. Names and locations of the orders are shown in Table 14.2.

At one time, there were as many as 92 federal milk orders. Due to new technology in highway construction, packaging of milk and dairy products, and factors affecting keeping quality of milk, marketing areas over which milk moved expanded greatly. This caused overlapping of marketing areas and brought about merging and consolidation of market orders. There are very few, if any, cases in which a federal milk marketing order, once operative, has been terminated for any reason.

Commodities Eligible for Coverage by Federal Marketing Orders

Almost all agricultural commodities produced in the United States, except the feed grains, are eligible for coverage under federal marketing orders under provisions of the Agricultural Marketing Agreement Act of 1937. Milk, fruit, and vegetable producers, however, have made the greatest use of this legislation in marketing their product.

What They Do—Their Major Provisions

In the case of fruits and vegetables, three broad categories of activities

TABLE 14.2.
Federal Milk Marketing Areas, United States, January 1985

North Atlantic	East South Central
New England	Tennessee Valley
New York–New Jersey	Nashville
Middle Atlantic	Paducah
South Atlantic	Memphis
Georgia	West South Central
Alabama–West Florida	Central Arkansas
Upper Florida	Fort Smith
Tampa Bay	Southwest Plains
Southeastern Florida	Texas Panhandle
East North Central	Lubbock–Plainview
Michigan Upper Peninsula	Texas
Southern Michigan	Greater Louisiana
Eastern Ohio–Western Pennsylvania	New Orleans–Mississippi
Ohio Valley	Mountain
Indiana	Eastern Colorado
Chicago Regional	Western Colorado
Central Illinois	Southwestern Idaho–Eastern Oregon
Southern Illinois	Great Basin
Louisville–Lexington–Evanston	Lake Mead
West North Central	Central Arizona
Upper Midwest	Rio Grande Valley
Eastern South Dakota	Pacific
Black Hills	Puget Sound–Inland
Iowa	Oregon–Washington
Nebraska–Western Iowa	
Greater Kansas City	
St. Louis–Ozarks	

Source: Federal Milk Order Market Statistics, U.S. Department of Agriculture, Agricultural Marketing Service, Dairy Division, March 1985, p. 9.

have been undertaken under federal marketing order programs. These are quality control, quantity control, and market support.

Quality Control

Quality control provisions are implemented through shipping restrictions on certain sizes and grades of the product. These provisions permit the setting of minimum grades, sizes, and maturity standards. These standards are usually enforced through mandatory federal inspection.

The rationale for quality control provisions being included in an order has two facets. Removal of off-grade products improves the average quality of the product that goes to the market. Such higher-quality products should, it is presumed, be more acceptable to the consumers and will probably command a higher price and larger returns to the producer. Since the use of quality control measures reduces the quantity of product available for sale in the short run, such controls may also be viewed as an indirect means of quantity control.

The use of quality controls varies considerably among orders and sometimes varies within the same order over time. Standards sometimes remain unchanged over several marketing years in an attempt to impose and maintain minimum levels of product quality which may be placed on the market. In those cases where quality standards are frequently changed, even within shipping seasons, this may suggest a concern with quantity and the use of quality standards in an attempt to control quantity. Quality standards, it is clear, are flexible and may be used for a number of purposes.

All but three fruit and vegetable marketing orders in effect in 1981 had some sort of quality control provisions. Of the orders, 37 permitted the use of both size and grade regulations, while a few permitted only grade or size regulations. Cranberry and cherry orders permitted grade and size standards for a portion of the crop. Florida grapefruit marketing orders do not permit size and grade standards, but the fruit sold under these orders is subject to such standards under the Florida citrus order.

The Agricultural Marketing Agreement Act of 1937 states that if certain specified commodities are covered by a marketing order containing quality, size, and maturity control provisions, any imports of these commodities must meet the same or comparable standards.

Quantity Controls

Quantity control provisions of such orders represent the strongest form of regulation under federal marketing orders. This form of regulation, of course, has the greatest potential of affecting price. The methods usually used are volume or sales management and market flow regulations.

These strategies are separate and distinct, but the objective in each case is to obtain a higher price for the commodity than might be received in the absence of a federal marketing order. Volume management provisions attempt to influence price by reducing quantity sold on the primary market. Market flow regulations, on the other hand, attempt to regulate the within-season pattern of sales in the primary market rather than control the total quantity sold.

The three methods used under the Act for volume management are producer allotments, market allocation, and reserve pools. These may be used singly or simultaneously under a single federal marketing order.

Under a producer allotment arrangement, a producer is assigned a maximum quantity that may be sold off the farm. This allotment is usually based on sales during some specified period or base. The total quantity to be sold is established each season and is given to the producer in the form of a percentage of the total allotments for all producers as applied to the individual producer base. For example, if the market order administration determines that 80% of the total base allotment will be sold on the basis of expected market conditions, then each producer's allotment will be 80% of the base.

Producer allotments have been used only for hops, spearmint oil, and Florida celery. A reserve pool is also used in the hops order to handle any excess production. In the case of spearmint oil, about 80% of U.S. production is covered, and imports are not important as in the case of hops. New entry into production is not barred. Florida celery allotments have historically been set far in excess of actual shipments. Specific percentage allotments are set for new celery producers as well as for those producers who have bases and who wish to increase their sales.

A market allocation program administratively dictates maximum sales in one of two or more different market outlets for the same basic commodity. Prior to harvest, a free or saleable percentage is determined based on the expected crop size and market conditions. Each handler then applies this percentage to the total quantity handled to determine the quantity that may be marketed without restriction. Sales in excess of the free or salable percentage must be in noncompetitive market outlets such as export, manufactured products, oil, or livestock feed. Free percentages may be increased during the season if the primary demand turns out to be better than had been expected, but they may not be lowered. Cranberries, almonds, filberts, California dates, and raisins authorize market allocations.

Reserve pool programs, as a control mechanism, are similar to market allocation programs, but differ in that the restricted portions are held as a set aside or reserve pool rather than diverted immediately to secondary markets. Sales can be made from the reserve pool on the primary market if demand conditions improve or if supplies prove to be smaller than expected. Such supplies may also be sold in primary markets in later years, diverted to secondary markets, or disposed of in nonfood uses. Tart cherries, spearmint oil, almonds, walnuts, filberts, raisins, hops, and prunes make use of reserve pools.

Market flow regulations is a form of quantity control. All of the product is destined for sale, but the amount sold each week during the shipping season is regulated to avoid seasonal gluts and shortages and the related low and high prices. These regulations are implemented through handler prorates and shipping holidays.

Handler prorates specify the maximum quantity a handler may ship in a stated period of time, usually a week. Product received in excess of this quantity must be held for shipment in subsequent periods or diverted to secondary markets. Citrus fruits, Tokay grapes, Florida celery, and South Texas lettuce use some sort of market prorate arrangement.

Shipping holidays provide a second means of regulating within-season shipments. This is a period of time in which all commercial shipments are prohibited. Orders specify the conditions under which holidays may be declared, the maximum length of the holiday, and the minimum period between holidays. In reality, this is a weak form of controlling market flow, and shipping holidays are usually limited to periods surrounding calendar holidays. This practice prevents a buildup of supplies in terminal markets during periods of restricted trade activity such as during a calendar holiday. Several marketing orders authorize such holidays as a means of controlling market flow.

Market Support Activities

Another category of provisions authorized under the Act is called market support activities. These do not directly affect quantity sold, but are aimed at contributing to the overall goals of the legislation as related to the notion of orderly marketing.

Standardization of containers and packs may be used to assure or promote greater uniformity in packaging. Handlers may be assessed through orders to raise funds to support research and also in the case of specific commodities for promotion. Handlers may be required to post minimum prices, and unfair trade practices may be prohibited. Shipping information is required of handlers which is necessary in the administration of orders.

Such market support activities are widely used in all of the major fruit and vegetable commodity groups. Many use the provision relating to research, and the Idaho–Oregon onion and the Florida celery orders permit advertising. All orders provide information regarding the commodity which is necessary in their administering. This information is aggregated to provide data that are useful in marketing decisions and is made public through the marketing order administrative process.

Market Order Administration

Federal market orders for all commodities except in the case of milk are administered by committees composed of representatives of growers and handlers with the counsel of U.S. Department of Agriculture personnel. The Secretary of Agriculture has final authority and issues regulations concerning the operation of marketing the orders.

It is worthy of note at this point that price determination or a method of determining prices is not permitted under federal marketing orders for these commodities. Certain provisions such as allotments, prorates, and reserves are permitted in orders in an effort to bring about a price which, in the judgment of the administrative board, will result in appropriate returns, but prices themselves may not be set. This is in contrast to the situation in federal marketing orders for milk, which we now consider.

Federal Marketing Orders for Milk

As previously indicated, there are 44 federal marketing orders for milk in the United States. These orders directly regulate over 70% of the milk sold in the United States and more than 80% of all the milk sold which is of fluid quality.

How Different from Fruit and Vegetable Orders

Three areas of difference stand out when provisions of the 1937 Agricultural Marketing Agreement Act as they relate to milk orders and to orders for other commodities are examined. These relate to quality control, pricing provisions, and to order administration.

As seen before, market orders for fruits and vegetables may make use of several provisions such as minimum grades, sizes, and maturity standards which are aimed at exercising control over the quality of the product sent to the market. In the case of milk, no control of these types is exercised under the order. Only milk of fluid quality as contrasted with that of manufacturing quality is accepted for regulation under federal milk marketing orders. Regulations regarding conditions under which milk is produced and handled on the farm, in the process of hauling, at the processing plant, and so on in order to meet fluid standards, standards permitting it to be sold as fluid or bottled milk, are established and monitored by the various states, with federal standards being involved with milk moving in interstate commerce. Once the milk is certified as meeting so-called Grade A or fluid milk standards, it is eligible for handling under the provisions of a federal marketing order. If it does not meet those standards, it is precluded from such orders.

Price Determination

The second difference or distinctive feature of federal milk marketing orders relates to price. In the case of fruit and vegetable orders, it is recalled that the administrative board makes a determination as to what would be an appropriate price to be received by the producer for the product. It then used, if necessary, certain provisions permitted under the Act such as allotments and prorates which, in the light of expected market conditions, would result in the desired price. Price itself was not established.

In the case of milk, federal orders are required to use a method of determining price, and this becomes an integral part of the order. Two general formulas are used in pricing milk under federal orders, both based upon the pricing of surplus milk, milk not needed for bottling purposes. Under one method, not directly used in most cases today, the market quotations of end products and yields of product are used to determine a gross value of, say, 100 pounds of milk. A charge for processing the milk into the products is deducted from the gross value and that becomes the price of surplus milk.

Another method, the famous M-W series pricing method, is based upon an average of prices paid by processors not regulated under a federal order for milk which they use in manufactured dairy products such as butter and cheese. This provides a basic formula mechanism which is used for the price paid for milk used in the lowest class use. Differentials added to this price, as presented and accepted in federal order hearings, are added to this price to establish prices for other use classes of milk. In this way, a complete pricing structure for all classes of milk under the classified pricing system is established.

Milk Order Administration

The third area of difference between provisions in the Act relating to milk and other commodity orders is in administration. It is recalled that fruit and vegetable orders are administered by a board made up of representatives of growers and handlers of the commodity, with the Secretary of Agriculture having final authority. Milk marketing orders, on the other hand, are administered by a market administrator who is appointed by the Secretary of Agriculture. The market administrator is completely responsible for carrying out the provisions specified in the designated market order.

What Milk Orders Do—Their Major Provisions

Federal milk marketing orders, as with orders for other commodities, operate under the provisions of the 1937 Agricultural Marketing Agreement Act. That Act, as indicated, was designed as a means of providing the complementary reinforcement to the farm sector of the dairy industry which it was not able to do alone. Stability and orderliness were deemed to be in the best interest of all. Milk orders constitute the program form designed to carry out the policy set forth in the Act.

Orderly marketing is encouraged by the use of classified pricing which had been developed and tried by dairymen, but which failed to contribute to its potential toward orderliness because of the free rider problem. The same is true of pooling and developing a means by which dairymen could develop a commitment to providing the consumption needs of a market area with an adequate supply upon which consumers could depend. Assurance and commitment of dairymen to this type of marketing arrangement would make it possible for them to commit adequate resources to the production of milk such that an adequate supply could, in fact, be assured. All in all, it suggested a situation of orderliness which was felt to be in the overall public interest. It is to this end that the provisions of federal milk marketing orders are designed to contribute.

How Are Orders Amended or Changed?

Federal milk marketing orders are amended or changed by using the same public hearing process as was used in starting the order in the beginning. The order was begun after certain groups asked for a public hearing to consider the use of an order to overcome problems that were bringing about instability and a lack of orderliness in the market. The hearing was held, evidence was submitted regarding the cause(s) of the problem and the provisions an order should contain in order to effectively deal with the problem. An order was established based upon the evidence.

It is logical to assume that as time and circumstances change, the provisions suggested for use in the order in the beginning may become less and less effective in performing as they were intended. For example, the location differentials in an order are tied closely to the cost of hauling milk. As fuel costs increased over the past few years, the location differentials based upon transportation fuel costs at that time were no longer appropriate. Milk may move in the wrong direction or not at all if the differentials are not in line with transportation costs. A movement back to a state of disorderliness begins to take place.

Interested groups, sensing the difficulty, will petition the Secretary of Agriculture for a public hearing to consider changing the location differentials in the order to reflect changes in hauling costs due to higher expenditure for fuel. A hearing would be held in exactly the same manner as before, evidence would be submitted regarding fuel costs, etc., and if the evidence warrants, the provision would be changed—the order would be amended.

Who Pays the Costs of Running an Order?

As previously indicated, all marketing orders involve monetary costs which must be covered. There are costs of immediate administration of the orders and for the role of the U.S. Department of Agriculture in administering them. There are also compliance costs which, in many cases, consist to a large extent of a paperwork burden of keeping the necessary data and information for compliance with the order.

There are handler assessments for order administration of both milk and other orders. U.S. Department of Agriculture costs for their part in administering the orders are covered in their regular pay scales and procedures. Compliance costs are difficult to estimate, but it is likely that most of the information required of growers and handlers for order administration is routinely generated through customary business procedures and is not an added expense.

What Is the Track Record of Orders to Date?

Criteria or concepts discussed earlier and that, by implication at least, were suggested as criteria against which the success or failure of federal marketing orders might be measured included orderliness in the process and costs of resource adjustment, the concept of price administration by a third party, and the social criterion of the public interest.

As is true in trying to quantitatively measure against criteria of this general nature, there are no easy answers. As measured against the chaotic conditions that existed prior to enactment of the 1937 Marketing Act, it would appear that the federal marketing order system has worked very well.

Information and Involvement

One area that stands out as a very positive outgrowth of the use of federal orders is that of information and involvement. It is generally agreed that information generated by the orders and its being made available as a public good in a general form has contributed to market orderliness. Producers, handlers, consumers, others in the linkages from producer to consumer, and attorneys have become involved in federal order procedures and in the process have become much more knowledgeable, not only about the workings of the orders themselves, but about their background, their economic and legal underpinning, and the whole area of agriculture. This is positive.

Income Distribution Effects

There probably has been some income distribution effects with flow directions changing from time to time, but these would be hard to measure and the overall impact would be difficult to assess.

Size of Farm

The preservation of the Jeffersonian concept of the small or family farm has been advanced as a goal of agricultural policy. On balance, it is felt that federal marketing orders tend to preserve existing farm structure.

THE ULTIMATE PERFORMANCE CRITERION

As usual in dealing with important questions such as the ones being considered here, positions will be taken on both sides of the issues. On balance, however, we will perhaps agree that the ultimate criterion relates to whether the public interest has been served by using federal marketing orders. A back door approach to this question would be to review the chaotic conditions that prevailed prior to passage of the 1937 Agricultural Marketing Agreement Act providing for federal marketing orders. In doing this, one can quickly come to the conclusion that those conditions were not in the interest of the general public. To the extent that orderliness has been restored and conditions are more equitable among all factors in a given market and between markets, it would appear that marketing orders have gone far in the direction of that golden mean between rigorous government control and tooth and claw mayhem which John D. Black called "assisted laissez-faire." They were dedicated in the beginning to the process of promoting the public interest, and it appears that they have not lost sight of that important role. Their kinship with cooperatives in the areas of historical development, their legislative underpinning, and their group action focus have equipped them well to serve as a supplement to action under the Capper–Volstead Act.

REFERENCES

Armbruster, W. J., Henderson, D. R., and Knutson, R. D. (Editors). 1983. *Federal Marketing Programs in Agriculture—Issues and Options.* Interstate Printers and Publishers, Inc., Danville, IL.

Deiter, R. E., Williams, S. W., and Groebele, J. W. 1983. Cooperative Activities in Marketing Fluid Milk in the Chicago Federal Order Market. Research Bulletin 597. Agriculture and Home Economics Experiment Station, Ames, IA.

Heifner, R., Jesse, E., Armbruster, W., Nelson, G., and Shafer, C. 1981. A Review of Federal Marketing Orders for Fruits, Vegetables, and Specialty Crops—Economic Efficiency and Welfare Implications. Agricultural Economics Report No. *477*. U.S. Department of Agriculture, Agricultural Marketing Service.

McBride, G., and Boynton, R. D. 1977. Class I Milk Pricing in Federal Order Markets. Research Report No. *334*. Agricultural Experiment Station, Michigan State University, East Lansing.

Nourse, E. G. 1962. Report to the Secretary of Agriculture, by the Federal Milk Order Study Committee. U.S. Department of Agriculture.

U.S. Department of Agriculture, AMS. Federal Milk Order Market Statistics. Published monthly with annual summaries.

U.S. Department of Agriculture, AMS. 1975. Questions and Answers on Federal Milk Marketing Orders. AMS No. *559*.

U.S. Department of Agriculture, CMS. 1971. Cherry Marketing Order for Cherries Grown in Michigan, New York, Wisconsin, Pennsylvania, Ohio, Virginia, West Virginia, and Maryland. Federal Register. January 23, 1971.

U.S. Department of Agriculture, CMS. 1971. Compilation of Agricultural Marketing Agreement Act of 1937. Agricultural Handbook No. *421*.

U.S. Department of Agriculture, FCS. 1971. Cooperative Bargaining Developments in the Dairy Industry, 1960–70, with Emphasis on the Central United States. Farmer Cooperative Service Research Report No. *19*.

U.S. Government Printing Office. 1983. Code of Federal Regulations, Agriculture. CFR 900.1, Parts 900 to 999. Revised as of January 1, 1983.

TO HELP IN LEARNING

1. What Federal marketing orders are being operated in your state?

2. Prepare a short questionnaire to use in measuring the knowledge about and understanding of federal marketing orders of five of your peers.

3. Arrange a personal interview with the market administrator of a federal milk marketing order. List several questions you wish to discuss (may be done by telephone).

4. Arrange a meeting with a member of the administrative board of a fruit marketing order. Discuss what the board does, how they do it, and other relevant questions.

5. Prepare a page of results of your findings in questions (2)–(4) and your conclusions from the findings.

DIRECT QUESTIONS

1. What is a federal marketing order?

2. Why are they in use?

3. How do you get one?

4. What is a federal order hearing?

5. What is the Federal Register?

6. How are federal marketing orders administered?

7. For what agricultural commodities can federal marketing orders be used?

8. How many federal marketing orders are in operation now?

9. There are fewer federal milk marketing orders today than 10 years ago. Why?

10. Who is regulated under a federal milk marketing order?

11. How do you change or amend a federal order?

12. Who votes in a referendum to decide whether a federal order will be used?

13. What does the market administrator of a federal milk marketing order do?

14. What is meant by orderly marketing?

15. What is meant by the public interest?

16. What is the M-W series?

17. What is classified pricing of milk?

18. What is a "shipping holiday" and where, when, and why are they used?

19. What is meant by "assisted laissez-faire"?

20. How often are federal order hearings held?

TYING-TOGETHER QUESTIONS

1. Prices to farmers were disastrously low in the early 1930s. Why?

2. What is meant by economic orderliness and how are structure and other concepts which we've covered in this book related to it?

3. Administered prices were discussed in this chapter. What does the term mean and how does the concept tie in with our objectives in this course?

4. Discuss economic adjustment in the context of the competitive model and in that of "assisted laissez-faire."

5. Marketing orders and agricultural cooperatives are said to be very complementary in their workings. Discuss in relation to our objectives in this book.

15

Cooperatives and Commodity Market Pools

Pooling is another activity which is peculiarly cooperative in nature. It is done by a group of producers for a common purpose—individuals do not engage in this activity. It reflects the amount of patronage of each producer included in the pool. It is democratically run, since pools are usually run by cooperatives for their members and are, in most cases, subject to the one person–one vote principle. Costs are shared on the basis of the contribution of each, and the costs and returns are based upon the patronage of each as reflected by the amount of product of each producer in the pool.

This kind of activity is not specifically mentioned in the Capper–Volstead Act or other cooperative legislation but, as said before, it is so cooperative in nature that it is almost taken for granted that cooperatives will engage in some kind of pooling. There is the general authorization in the legislation, of course, to engage in marketing activities and to provide marketing services to members, and this may be interpreted as an implicit sanction. Despite this, a cooperative must make specific provisions for engaging in pooling in its bylaws and must strictly adhere to such provisions. Provisions for pooling can also be made in the cooperative's marketing contract with its members, but the use of a contract itself should be provided for in the cooperative's bylaws.

Pooling, or the operation of commodity market pools, is a business arrangement which, as has been indicated, is uniquely cooperative in its nature. It is also uniquely associated with the nature of agriculture in that agricultural commodities are handled in their raw state directly from the producer and lend themselves to comingling.

Further, it is uniquely cooperative in its underpinning in that it permits a group of individual farmers to combine their offerings into larger lots with potential benefits for all concerned. Greater uniformity with respect to grades and standards is possible with larger lots. This is very much in keeping with the movement toward specification buying and full supply contractual arrangements which are becoming more and more commonplace in today's marketing.

WHAT IS A MARKET POOL?

As indicated previously, the products from the farms of many producers are combined into a market pool for sale to various buyers. Proceeds from the sale of the product are divided among the pool members after transaction costs or expenses have been deducted. Each member of the pool receives the average price for each unit of commodity contributed to the pool. Provision can be made for differences in quality, transportation costs, or services rendered by the pool or by the producer. Once the proceeds are determined and the agreed upon costs are calculated, each producer whose commodities are in the pool receives the average or pool blend price for the products.

HOW POOLING DIFFERS FROM
OTHER MARKETING METHODS

In other marketing methods such as the buy–sell operation, the producer maintains ownership rights to the product until a price and other terms of trade are mutually agreed upon. Once agreement is reached, the producer receives payment in full for the product unless some other arrangement such as a price later plan or other pricing basis is used.

Under market pooling, producers turn over the pricing and marketing decisions to the cooperative marketing staff and agree to accept the average price for the pool after adjustments for cost, quality, and any other differences have been made. In most cases, the producer receives an advance payment when the commodity is delivered or at some specified time. The amount of the advance and its timing are determined on the basis of agreement and understanding as to procedures used by the cooperative. If the pool contents are sold over time, the producer may receive progress payments as pool sales progress. After the entire pool has been sold, all operating expenses, capital retains, and any other similar items are deducted and the producer receives a final payment. It is here that final adjustments for quality, grades, standards, and the like are made.

KINDS OF POOLS

There are two main types of pools: seasonal and contract. The basic difference in the two relates to the degree of control over price which is retained by the producers of the pooled commodity.

In a seasonal pool, by far the more common, the producer agrees to deliver some specified portion of a crop to the cooperative and to accept the pool price which has been adjusted for all specified costs. The producer, in this case, is turning the product over to the cooperative to be comingled with the offerings of many other members. Professional marketers at the cooperative handle the marketing. The pool is calculated and the producer–member receives the pool price for the product. This plan is used by many dairy, rice, fruit, and vegetable cooperatives.

The contract pool is of two general types—the call pool and the purchase pool. As indicated previously, the basic difference in these types of pools is in the degree of control retained by the producer over the price and perhaps other terms of trade regarding the product offered.

In a call pool, the producer sets a minimum or reservation price below which the commodity may not be sold. Delivery of the amount of the commodity which is committed is usually made before some date that was fixed early in the pooling period. In the usual purchase pool, the producer exercises whatever price-determining power that is possible by timing of delivery to the pool. The price received is usually the expected cash price on the day of delivery.

A contract pool, while being referred to as a pool and having many of the characteristics of a pool, is not a true pool at all. In such pools, prices are determined on an individual basis, as is the case in buy–sell types of transactions. Such pools are, in effect, marketing agreements that make it possible to pool expenses once individual prices have been established.

It is apparent that this type of pool would not fit very well with cooperatives.

Pooling arrangements can be very flexible and can take on many forms. They can vary in form with respect to duration of the pool, the number of commodities in the pool, how grades and/or standards are handled in the pool, and any other unique characteristics of the commodity or in its production.

WHY USE POOLING?

Several reasons can be given for using a pooling plan in marketing agricultural commodities. These include the following:

1. Possibly higher prices and/or better terms of trade for the members' products.
2. Spreading and reducing price risks.
3. Specialized efforts by marketing specialists.
4. Greater orderliness and stability in the marketing of the product.
5. More complete control of quality.
6. Complementary with movement to specification buying.
7. May reduce transaction costs.
8. Serves as an educational marketing tool.
9. Promotes cooperative ideal of unity.

Let's discuss each of these possible advantages.

BETTER PRICE AND TERMS OF TRADE

Most cooperative members and leaders, when asked why their cooperative was formed, will not hesitate to place getting a better price for their product at or very near the top of their list of reasons. Other conditions or terms of trade which may or may not translate into a higher per unit price will very often be a part of the explanation. In one case, the history of the cooperative reveals that a major impetus toward beginning the cooperative was the fact that buyers of their product would not return the empty containers that were used to ship the product in time for them to be used for the next shipment. This necessitated the purchase of two sets of the containers, an increase in their costs, and thus, in effect, a lower price.

There is, of course, no guarantee of the price received by using a pool, but such an arrangement should, in most cases, provide a better price for its members than would be the case otherwise. The greater intensity of the effort put forth by the marketing specialists operating the pool should make it possible to have more knowledge of the market and better interpretative ability in regard to market data. This should result in better timing and pricing of sales and an increased ability to take advantage of all marketing opportunities that present themselves.

The marketing specialists' hands are strengthened by having access to more complete market information and the ability and time to interpret, analyze, and utilize it. They will also be helped by having much larger quantities of the commodity readily available than is usually the case. They can exercise much greater control over quality, grades, and standards and the resultant uniformity than would otherwise be the case. The element of greater control in terms of quantity, quality, timing, and so on may translate, of course, into increased market power from which better prices or terms of trade are exacted. There may be an element of power involved but, in many cases, large, specialized buyers of a commodity stand ready to pay a premium for the assurances involved in larger quantities and uniform standards and quality. These can be provided under a pooling arrangement, and this would not be the case in other arrangements.

SPREADING PRICE RISK

The second possible advantage in using a pooling arrangement, that of spreading and reducing price risk, is closely tied in with the points previously discussed. Higher prices may be possible for the reasons indicated, but regardless of the price level, all pool members receive the same average price for their product, and the risk of some receiving a lower than average price is eliminated. In the same way, losses are spread among all pool members. This, of course, is the essence of cooperation.

SPECIALIZED SELLING EFFORT

Reference has already been made to the expectation that specialized effort on the part of marketing specialists should result in sounder selling decisions which, in turn, translate into better prices and/or terms of trade. An advantage of specialization relates to the opportunity to give undivided attention to the task at hand—that of selling the product under the best terms. Marketing conditions can be studied, trends noted, developments in direct and perhaps indirect areas can be kept at hand, and their implications taken into account. The needs and peculiarities of the buyers and the potential buyers of the product can be constantly kept in mind, and supplies can be matched and coordinated with them. A general growth and development in the selling expertise involved can result in a selling activity attuned to most elements which, if properly taken into account, may result in benefits to pool members in the form of better prices and terms of trade and to buyers of the pooled products in the form of commodities that more nearly match their needs in quality, quantity, and other relevant dimensions.

ORDERLINESS AND STABILITY

Another by-product of pooling operations which can be beneficial to both sides of the supply–demand equation is the greater orderliness and stability in selling the product. These benefits stem, of course, from the control of the quantity, quality, and timing of the pool operations provided for by the cooperatives' bylaws and spelled out in the marketing agreement between the member and the cooperative. Flexibility in directing the commodity in appropriate amounts of desired qualities is made possible under pooling arrangements. The timing of the movements, the actual movement, the point to which the product should be moved, and the shipping or movement conditions with respect to quantity and quality control are all amenable to planning and orderliness in the process which, in most cases, is not possible otherwise. Deliveries can be tailored in grade, quantity, timing, and other conditions to meet the requirements of buyers, and markets being served in the most complete sense will inevitably result. Agreements can be made with potential buyers, and such buyers, realizing that security of expectations which can result from such orderliness and stability are beneficial to them, can contribute their part to the conditions which are in the best interest of all. Timing of commodity movements over a specified period can soften and perhaps eliminate temporary surpluses and gluts on the market. This is a stabilizing factor and contributes to market orderliness in that day-to-day price fluctuations to which farm commodities are subject may be reduced.

QUALITY CONTROL

Implicit in some of the benefits of pooling which have been mentioned is the part played by the possible greater control of quality.

Depending upon the provisions of the marketing agreement between the member and the cooperative, the control of quality may go as far as quality and variety specifications to the producer. These have become very common in the case of fruits, vegetables, and cooperatives handling similar commodities. It is understood that the pool is assuming responsibility for the commodity from the beginning, and this gives the pool selling specialists the ability to control the product's quality much more effectively. This decreases losses due to damage in transit and otherwise, and makes it possible for the cooperative to develop a reputation as a dependable supplier of a quality product that meets market demands.

MARKET SPECIFICATIONS

One of the most interesting and perhaps most significant movements in the product marketing area over the past decade has been that of specification buying and the resultant necessity of meeting the specifications that are laid down. This has been most evident in fruits and vegetables, but elements of the movement can be found in many commodity areas. The usual scenario is one in which the retailer—in most cases, the supermarket—specifies that it wants, for example, X tons of cucumbers, within a range of 3 to $3\frac{1}{2}$ inches long, and of a light green color delivered to its receiving dock with specified processing having been done by 8 o'clock each Friday morning. Other products such as sweet corn, tomatoes, and other fruits and vegetables are specified in a similar manner.

It can be easily seen that meeting these requirements calls for the ultimate in control of such areas as variety, timing, cultural practices, and harvesting methods. It can also be easily seen that the type of market system which developed to service the offerings of the thousands of individual farmers producing the products they wished to produce, in the way in which they wished to produce them, and harvesting them when they wished would no longer be adequate.

The coordination of effort which is essential to meet the needs of the new market demand can perhaps be provided only through contractual arrangements that spell out exactly what is required from the first step in planning and carrying out production all the way in each step to harvest. The role of the cooperative and pooling in such an effort is, of course, most obvious.

The marketing agreement between the member and the cooperative would specify the details in every step necessary to meet the specified market demand for the product. The pooling arrangement, along with its selling specialists who know what the market wants and when it is wanted and the production technologist capability inherent in the producer and some field service, is uniquely capable of meeting this marketing phenomenon. Rather than to appear to be balking at doing some of those things that are essential if the market requirements are to be met, the more logical posture is to accept that these are the requirements being laid down by the consumer and, in meeting them, the interests of all parties are being served. It drives home, with vigor and conviction, that production is not just for production's sake—it is for a market.

SELLING VS MARKETING

This reflects the need for a change in attitude and philosophy which has long prevailed on the part of producers of agricultural products. The change necessary is embodied in the business school concept of selling vs marketing. Marketing in this concept is centered on the position that agricultural producers have a product(s) and they would like to have someone buy it. Selling, on the other hand, is based on the position that we know what you want and we have a product that fits your wants, or we are tailoring products to fit your wants and needs. This is completely accepting the position that agricultural products are produced for consumers whose tastes and preferences and abilities to buy are different, perhaps, from those of the producers, but recognizing that as being of no importance. Tastes and preferences of the consumer are recognized as being the significant ones, and every effort is made to satisfy them with the foods and services they want. This involves market segmentation and tailoring products to fit each segment.

This suggests, as has been indicated, that there is no place for a production–marketing dichotomy which prevailed in the past and may still prevail to some extent. Specification buying, instead of being viewed as a problem, should be viewed as an opportunity to move wholeheartedly away from the marketing position to one in which selling is the driving force. Pooling, properly handled, has the potential of helping producers of agricultural products come to appreciate the importance of accepting a consumer rather than a production orientation. Larger lots, greater uniformity, timeliness, quality, and other factors involved in product specifications which are required to meet consumer preferences can easily be handled in pooling arrangements.

LOWER TRANSACTION COSTS

Pooling may also reduce transaction costs. The potential for doing this comes about in a number of ways. As compared with each individual producer dealing with potential buyers, the savings in costs may be substantial.

The expertise involved in having selling specialists who devote full time to gaining the knowledge necessary to become effective sellers can contribute to effectiveness in the marketing area from an efficiency standpoint as well as from the standpoint of gaining higher prices than would otherwise be possible. Such a staff would be completely attuned to market specifications, potential buyers, the process of finding new markets, etc. Having larger lots available of a known quality makes matching the volume and product specifications of buyers much easier and more efficient. The form, timing, and all logistical components of the selling process are areas that offer great potential in effecting savings and reducing transaction costs. All of these areas also contribute to greater stability and orderliness over time in the marketing process. This pays dividends in the form of reduced costs, not only to those who are most closely involved in the process, but it serves the interests of the ultimate consumers in that their needs and wants are more adequately met in an efficient manner. Improved quality and the greater quality control which is possible through the pooling process also effect lower transaction costs and serve the interests of consumers and, thus, those of the general public in a significant way. The potential of pooling in reducing transaction costs and making the selling process more effective in achieving its ends is great.

LEARNING FROM INVOLVEMENT

It has been said that one of the most effective teaching and educational tools is an involvement in the area about which greater understanding is needed. This is true in the case of pooling in the marketing process.

This does not mean that the members of the cooperative as producers would seek to become selling specialists in running the pool. It does not mean that they, as individuals, should strive to understand all the intricate details that are essential in the case of those who spend full time in studying market conditions and developments.

It does mean, however, that some understanding of the process is necessary if the pooling technique is to be used as effectively as it can be in meeting the needs of the market. The discipline necessary in meeting the quality, quantity, variety, timing, and logistical requirements of effective pooling can drive home the importance of the fact that agriculturalists are engaged in an activity in which selling and meeting the specifications of the consumers are dominant factors in the whole process.

IMPLICATIONS FOR COOPERATIVES

This, of course, places a special obligation on the part of cooperative management and the board of directors to include the whys, hows, and requirements for success in the pooling process in their education and training agenda.

The very essence of cooperation is unity. This unity refers to purpose for the most part. Unity of positions regarding methods may not be realized at all times, and it may be that this is not an essential or even desirable goal. Diversity of backgrounds and training should be conducive to bringing different points of view to bear on problems. These can be expressed in different settings, such as at local, district, and regional meetings, and can be productive in reaching a position in regard to actions that should be taken. Once this position is taken by the board of directors, a unified position in regard to making the program work in meeting the overall goals is important.

Since a basic goal of most cooperatives is one relating to better prices and better terms of trade for their products, and since pooling has the potential of contributing to the achievement of this end, it has the potential of promoting unity of purpose and greater cohesiveness among cooperative members. Joint selling efforts through their cooperative and through use of pooling has the potential of better prices and better terms. In addition, it has the potential of increasing the producers' awareness of their market interdependency, and this is conducive to greater unity. An end result may be that of providing greater future security of the family farm.

ARE THERE SOME CONS ABOUT POOLING?

As is true in most cases, there are elements involved in seasonal pooling which might be considered as being disadvantages or problems by some. Many of these stem from the rather strict requirements for successful pooling. These would be included among such considerations:

1. Opportunity for involvement in the selling process by the member is curtailed.
2. Depending upon the cooperative's selling strategy, some flexibility in selling may be forfeited.
3. Optimum conditions with respect to quantity and quality of the pool may be difficult to attain.
4. Full payment to members is delayed.
5. Complexity of the operation.
6. Change in concept of selling.
7. Members may not understand the cooperative's needs for capital funds and how they may be provided through the pooling mechanism.

Again, let's discuss each of the areas, which may be referred to as disadvantages or problems.

INVOLVEMENT

One of the areas in which agricultural cooperatives differ from other forms of business is that of the strategic role played by members as an integral part of the management trio. Effective performance of this role requires knowledgeable involvement of members in the affairs of the cooperative.

As was pointed out previously, however, in exercising that role, direct involvement in the everyday operations of the cooperative is left strictly to the manager in carrying out the policies established by the board of directors. The policy that is established by the board of directors and that is carried out by the manager through programs devised for this purpose should reflect the thinking of the members as they participate in the affairs of their cooperative at the local, district, state, and regional levels. It is here that opportunity is provided for discussion and debate in regard to the cooperative's objectives and how they might best be met. As the results of the discussions and suggestions move through this process, they may be formulated into resolutions for consideration of the voting delegates, or the entire membership in some cases, at their annual meeting. It is from these resolutions that the board of directors receives guidance from the members in regard to what they consider to be appropriate policy, with some indication, at times, as to programs that might be used to carry out the policy. In this manner, the member exercises a vital management role as a part of the management trio, and if such a role is based upon knowledge and information, the member is performing in a professional manner and is contributing to the effectiveness of the cooperative in achieving its goals.

Once this function has been appropriately performed by the member at all points and the results of the functioning have been formulated into policy by the board of directors, the member then accepts the policy. The manager devises programs for carrying out the policy. The members participate in the program because it reflects their own contribution to the cooperative policy-formulating procedure. An evaluative posture is assumed, of course, by perceptive members with the idea that the same type of procedures will be followed time after time, and if the present policy and the programs devised to effectuate it are not performing satisfactorily, opportunity will be provided to offer suggestions for changing or replacing them with what appears to be better policy and programs.

If a cooperative is using a seasonal market pool arrangement in attempting to achieve its goals, it is assumed that such an arrangement resulted from the process just described. In this case, for the individual member, regardless of how capable he or she may feel in selling products, this function is turned over completely to management through the selling specialists who operate the pool. Participating in the policy formulation process and participating in it once programs are being used to carry it out are the appropriate roles of the member. Preparing for future involvement in this distinctive role by observing and collecting relevant data and information in regard to present policy and programs in a professional manner represents the functioning of the management trio at its best. In no way can it be said that the individual members have sacrificed control over their selling. They chose to form a cooperative in the first place because they recognized the weaknesses of the individual producer in trying to carry out the selling process.

This discussion, again, points up the importance of the cooperative concerning itself continuously with carrying out efforts in the area of education for its members. The important and necessary function by the members in this process of policy formulation can be knowledgeably and effectively performed only with a complete understanding of what is involved, what the goals of the cooperatives are, the role which members are supposed to perform, and the rationale for activities being carried out.

LOSS OF FLEXIBILITY

In most cases, cooperative managers attempt to concern themselves with a sound, longer-term economic position of the cooperative. This may mean that what appears to be short-term opportunities for gains will have to be sacrificed if policy has reflected a concern with the longer term and programs have been put into place which are in keeping with this emphasis. In such cases, the programs may not be flexible enough to take advantage of the short-term situation that seemingly exists.

This brings up a number of points for consideration. In the policy formulation process and in the programs designed to carry it out, the relevant considerations involved in long- vs medium- and short-term operations should be thoroughly discussed. The cooperative itself, the reasons for its establishment, its goals, its products, and the environment within which it operates are all important. Once the relevant information is laid out and policy based upon it is established, then the die is cast for the time being. If that policy reflects a long-term orientation of the cooperative, there is no basis for suggesting that sacrifices are being made because of what appears to be short-term opportunities presenting themselves, and the cooperative has to pass them up. The policy was established on the basis of sound considerations as being the best alternative course of action, so it should be followed until there are sound reasons for change.

At the same time, some degree of flexibility may be built into a long-term plan of operation without sacrificing to an unacceptable extent the potential of the longer-term focus. It, at least, could be reviewed annually with respect to its performance, its flexibility, how much is deemed desirable, and other factors. The mechanism for changing the time focus and for building varying degrees of flexibility into any pooling plan is available for possible use. If judiciously used in the way in which it is intended, it can be effective in establishing policy adapted to almost any situation.

POOL SIZE AND QUALITY

There is always difficulty in meeting conditions regarding pool size and quality which are necessary at a given time to maximize returns to the members. Given the biological nature of agricultural production and the fact that there are conditions such as climate and seasons which are outside the control of the most capable selling specialists, it is reasonable to expect that the task will not be easy.

However, the judgment regarding the weight given to this difficulty in deciding whether to carry on pooling operations is balanced against the difficulty and problems involved when individual producers attempt to carry out their own selling operations. The cooperative itself was set up because of these difficulties as a means of overcoming them. Should a pooling operation be deemed appropriate as a mechanism in helping the cooperative to reach its goals for its members, despite the fact that to expect trouble-free operations with no difficulties whatsoever is not reasonable, then it should be undertaken.

FULL PAYMENT DELAYED

One of the major differences between a pooling operation and a buy–sell type of operation is that full payment is not received by the member until the end of the pooling period. In the case of buy–sell operations, the producer receives payment immediately once terms of trade are agreed upon. The delay in payment under the pooling process has been mentioned as a problem or disadvantage. Cash flow requirements and other needs of this nature may dictate the weight given to this concern.

Advance partial payments are usually involved in a pooling operation. The size and number of such advance payments will depend upon the financial condition of the cooperative. Should such payments be large enough, the cash flow problem may be held to a minimum. The economic benefits of the pooling process have been shown to reflect higher than average prices being received in most cases by the members. Whether this is sufficient to offset the delay in payment may, of course, vary with individual members. There is no way that a pooling operation can operate on a buy–sell basis. This means that the member input into the management process, as has been suggested, should take into account all relevant information, including the need for immediate payment for products vs delayed payments, despite the fact that such payment may be larger.

POOLING IS HARD TO UNDERSTAND

One argument against pooling is that it is complicated and hard to understand. Its complexity is tied in with the fact that the meshing of a great number of factors is necessary if it is to be successful. Quality and quantity of a product tailored to known market outlets at specified times are essential if the process is to be successful. Perhaps these rather strict requirements for success are interpreted as complexities, and they do, in fact, complicate the process. Equitable allocation of costs and returns may further complicate the procedure, especially in situations in which a cooperative also operates on a buy–sell basis for some commodities. Other such circumstances may also contribute to the complexity.

Again, it seems that we must revert to the role of the member being involved in a knowledgeable manner in the prescribed role in the affairs of the cooperative. It also stresses the importance of the cooperative itself, taking seriously its obligation to provide relevant and timely education to its members in this as in other areas.

The goals and objectives of the cooperative in meeting the needs of its members must be kept in mind. Methods or alternative courses of action which may be used or taken in reaching those goals should be thoroughly researched and examined. The relevant factors for a particular cooperative should be taken into account. The problems and potential complications and complexities of operating a market pooling operation, as compared with other methods of operations, should be fully aired. Should such an operation be undertaken without thoroughly covering all points in a critically constructive manner and the outcome being positive on balance, there may be little assurance that a pooling operation will effectively contribute to achievement of the cooperative's goals and objectives.

CHANGE IN CONCEPT OF SELLING

The fact that pooling involves a concept and philosophy of marketing which is different than that traditionally held by producers is sometimes mentioned as a disadvantage or a problem. It is different, of course, from a buy–sell type of operation, but the decision on the part of the members to abandon their own individual efforts to sell products and to become a member of a cooperative should, within itself, have signaled a willingness to adopt new selling procedures. Choosing to use a market pool arrangement by the cooperative may be termed a change in degree rather than a complete change in marketing philosophy on the part of cooperative members.

Viewed in this manner, the use of a market pool arrangement as a selling strategy in the cooperative's attempt to meet its objectives may be considered an advantage rather than a disadvantage. This is especially true if the decision to adopt such a selling strategy was based upon the proper functioning of the peculiarly cooperative role of the members in exercising their management function.

The focus on selling rather than the traditional posture relating to marketing can be cited as a definite advantage in using a pooling arrangement. Determining potential buyers of the product, with that potentiality being based upon the fact that the product meets specified requirements in all dimensions with those products that are in the pooling arrangement, completes the movement from a marketing to a selling posture. This movement under no circumstances could be termed a disadvantage.

EQUITY CAPITAL
NEEDED BY COOPERATIVES

As has been pointed out previously, cooperatives, just as any other form of business, must have capital funds if they are to carry out their functions. The cooperative has many needs for tools which are necessary in its operation, and unless those needs are met, the cooperative cannot be successful in meeting its goals and objectives.

A pooling arrangement can be used as a mechanism for meeting the cooperative's needs for funds for capital improvements or for other purposes. This is usually done by retaining a certain amount or a certain percentage of the proceeds received from the pool by each member. The position taken by those that this is a disadvantage or a problem in pooling stems from the possibility that the member may not understand the need of the cooperative for capital funds and may not view such retains or withholdings with favor.

Again, this argument or position is improperly directed and is no argument against the use of pooling. Instead, it further emphasizes the importance of the cooperative's educational efforts with its members and in making sure that the members perform their unique role in cooperative management in the way that they must if the cooperative is to succeed in its responsibilities and obligations. In short, the argument that a lack of understanding is a fault of the market pooling arrangement is misplaced. Any lack of understanding on the part of members regarding the fact that equity capital funds are absolutely essential if their cooperative is to perform satisfactorily rests squarely on the importance attached to the cooperative educating members and the members' performance in carrying out their roles as a part of the cooperative management team.

POOLING AGREEMENTS

As has been indicated, success and effectiveness of a market pooling arrangement are directly dependent upon a number of requirements.

Assurance of a pool of adequate size and quality at an appropriate time is essential. Once the specifications or requirements which the buyer has placed on the product are met, the characteristics of the products in the pool must match the specifications or requirements. This matching calls for specialized selling expertise on the part of those operating the pool and discipline on the part of the producer–member whose product is to be included in the pool. If time, quality, quantity, and other specifications are to be met, a great deal of discipline on the part of the producer is necessary. Exacting product specifications are usually not met just by chance. As an aid in increasing the probability that such specifications can be met in a fairly orderly manner, a pooling agreement or a marketing agreement is used by a cooperative with its members. Such agreements may vary in content, of course, but basic areas covered outline the rights and responsibilities of both the cooperative and the member, the parties to the agreement.

THE PRODUCER–MEMBER MAKES A COMMITMENT

The marketing or pooling agreement on the part of the producer–member of the cooperative is that a specified portion of the crop or product being grown or produced will be delivered to the pool. The commitment may be stated as a percentage of the amount grown or produced by the member, a given volume, or in some cases, the amount produced on a specified number of acres. In some cases, variety and grade specifications are included as are cultural practices, harvesting methods, and timing. In the case of milk, grade requirements are always specified. Storage methods may also be specified.

THE COOPERATIVE MAKES A COMMITMENT

The cooperative commits itself to taking steps to assure that the best price and terms of trade will be received by the member for the products committed to the pool. There is also the commitment to do this as efficiently as possible with the largest net proceeds possible being remitted to the members whose product is pooled. The pooling agreement usually grants the cooperative the authority to establish practices such as grading, classification, handling, financing, storing, and testing, practices that will be helpful in assuring pooled products which meet market specifications.

HOW LONG DOES AN AGREEMENT LAST?

The length for which the agreement is in force is stated as a part of the agreement. This period of time is in keeping with the characteristics of the pool and the commodity being pooled. The manner in which early termination of the agreement may be allowed and the circumstances under which this may take place are usually specified. In the case of products such as milk and for which monthly pools may be operated, provision is for either party to terminate the agreement at a specified period during the year. If this is not done, the agreement is automatically extended. Provision for pool carryover and how the remaining product, if any, will be disposed of is made in the agreement.

HOW ARE PAYMENTS
TO PRODUCER–MEMBERS MADE?

In most pooling or marketing agreements, an advance payment is made to the producer–member at the time the product is committed to the pool. An in-progress payment may also be made at some time in the pooling period and then the final payment is made at the end of the pool period. The timing of each type of payment is specified in the agreement, as are the methods to be used in determining their size. Provision may be made that the pool has the right to the pool's contents as collateral to obtain funds for use in making the advance and progress payments.

HOW IS THE POOL FINANCED?

A definite method of allocating expenses among members and in deductions made for expenses will be set forth. Deductions for capital funds needed by the cooperative will be provided for in the agreement in most cases.

WHO IS IN THE POOL?

Pool members are usually qualified on the basis of the commodity being produced, type of operation, or location. Requirements for membership in the cooperative and the pool are specified in the membership agreement.

OTHER PROVISIONS IN THE AGREEMENT

Pooling agreements may also contain other provisions. These may include penalties for breach of agreement by either party, special or other limitations of the agreement, and conditions for the renewal.

POOLING IN KEEPING WITH COOPERATIVE PHILOSOPHY

As has been said, market pooling as a selling strategy is wholly consistent with the cooperative philosophy of group action in a concerted effort to solve problems. It is a device which is complementary to the cooperative method of operation in a very full and complete sense.

REFERENCES

Dunn, J. R., Thurston, S. K., and Farris, W. S. 1980. Some Answers to Questions about Commodity Market Pools. Circular *509*. Purdue University—Extension, Lafayette, IN.
U.S. Department of Agriculture, FCS. 1971. Implications in Coordinating Activities of Bargaining Associations. FCS Information *63*.

TO HELP IN LEARNING

1. Construct a five-question questionnaire and use it to determine the extent that your peers understand what is involved in agricultural commodity pooling. Survey three or four of your peers.

2. Tabulate the results of your survey. Set up a scale and rate the extent of knowledge indicated by your peers.

3. Prepare a report, based on the results of your survey, to be given to your peers who were surveyed showing them the results and what conclusions you reached regarding them.

4. If your findings indicated generally low ratings on your scale, what suggestions would you give to those who were in the survey? Why?

DIRECT QUESTIONS

1. What is pooling? From where does the name come?

2. How does it tie in with agricultural cooperative thinking and philosophy?

3. What is a buy–sell operation?

4. What is a seasonal pool?

5. What is a contract pool?

6. What is a call pool?

7. How might higher average prices be received under pooling than under individual selling of a commodity?

8. What is specification buying?

9. What is the difference in selling and marketing?

10. Who determines the specifications in market specification?

11. What is meant by market segmentation?

12. What is meant by the so-called production–marketing dichotomy?

13. What are transaction costs?

14. What, in your judgment, is the greatest disadvantage to the use of pooling?

TYING-TOGETHER QUESTIONS

1. Someone has said that cooperatives and pooling are not complementary in nature. Comment.

2. The comment has been made that there really is no difference in the concept of selling and that of marketing. Comment. State your position and why. What is the significance for agricultural cooperatives?

3. What, in your judgment, does the movement toward market specification portend for the future of agricultural cooperatives and their use of pooling as a marketing strategy or technique?

16

Member Education, Communications, and Cooperative Leadership Development

We have just finished discussions of such areas as cooperative financing, pooling, and marketing orders which are very important to the economic viability of a cooperative. They are also among the most tangible aspects of operating a cooperative in that capital requirements can be determined with a rather high degree of accuracy, and methods of assuring proportionate contributions to those needs can be found. Ways of distributing the accumulated equity which are consistent with cooperative principles are available.

We turn now to an area that is far less tangible than is financing or the other areas discussed. There is no way of knowing exactly how much is needed and how the burden of providing for these requirements will be borne in accordance with the proportionality principles when no specific equity accounts can be kept and when there is no place for distribution of such intangibles which may have been accumulated.

Despite our lack of ability to quantify what is involved in cooperative member education, communications, and leadership development, we are confident that they are just as important to the viability of a cooperative as is adequate financing or successful pooling, which can be quantified. As a matter of fact, they may be more important than the more tangible areas because without adequate member education and communications, the financing requirement may not be met over time, and without adequate leadership development the cooperative may be doomed to lose viability.

Two of the Rochdale Principles dealt with member education and information. These have been judged in modern times to lack the basic or fundamental nature of principles and are relegated to the status of practices. They are, nevertheless, considered essential to healthy cooperative performance. It is now fairly widely accepted that cooperatives overlook member education, communications, and leadership development programs only at their peril. They, or their importance, can be quantified only perhaps through proxy measures such as whether adequate financing is being realized, but they are extremely important. Their essential nature is recognized.

PROXY FOR MEASUREMENT OF ADEQUACY IN THESE AREAS

We spoke of the difficulty in measuring in a tangible manner whether efforts in member education, communications within the cooperative, and/or cooperative leadership development were adequate. Are they effective? Are all parts of the effort contributing to the objective(s) in these areas? Are there lessons to be learned in making such efforts more effective?

We suggested that if the cooperative was being adequately financed and if satisfactory equity redemption plans had been devised and implemented, member education, communications, and leadership development were probably adequate and effective. It appears unlikely, however, that without explicit recognition of the importance of these areas by the board of directors and manager as reflected in definite efforts as a part of the cooperative's activities that even the capital accumulation and equity redemption criteria can be used as a long-run measurement of success in these less tangible areas. It appears safe to assume that if members are not knowledgeably and meaningfully involved in their cooperative, then not only are the areas of communication, member education, and cooperative leadership development not being served, but other more quantifiable areas will eventually suffer.

The cooperative is a business organization designed to provide services to its members. This is implicit. At the same time, because of its service, use, and member orientation, there are certain requirements relating to membership involvement and member loyalty which are uniquely cooperative in nature. If this is not recognized and if the cooperative board and manager do not choose to take the tedious, painstaking, laborious route of involving members in a meaningful way in the cooperative, the likelihood of its becoming viable and remaining so over time is not great. Transaction costs are much higher in this sense than for private corporate arrangements, but any attempt to take shortcuts in this area is fraught with danger.

PROXIES

We've spoken previously of the difficulty in measuring the quantity and especially the quality and effectiveness of efforts made or resources used in areas such as member education, communications, and leadership development. Whether education as such, for example, can be measured or whether there are results or outcomes that may be related to educational efforts and that lend themselves more readily to measurement is our concern at this point. Is there something that adequately represents what is relevant in the three areas of education, communications, and leadership development which we have chosen as the title for this chapter? Maybe there is or maybe there isn't a proxy for all these, but let's use member involvement as one that may reasonably be used for this purpose.

MEMBER INVOLVEMENT

The rationale for using degree of member involvement or participation in a cooperative rests on the assumption that cooperative leaders have taken seriously the position that cooperatives have the duty to educate. For education to take place, some sort of communications process must have been used and parties must have communicated. If education had been achieved, it is further assumed that in the natural course of events potential leaders would emerge. All of this process would reflect member participation or involvement—thus, the use of member involvement as a proxy for the three areas with which we are concerned seems to be justified.

THE DUTY TO EDUCATE

We relegated member education to the status of a practice and not a cooperative principle in Chapter 6. There was the strong implication intended in doing this that it could very well be given a principle status and that cooperative leaders who do not engage in meaningful member education activities do so at great risk to their future viability. As previously stated, Abrahamsen has listed duty to educate as one of five cooperative principles which he deems to be basic in distinguishing cooperatives from other types of businesses.

Another one of his five principles is member control, which we discussed in our management triumvirate in Chapter 8. It would not be reasonable to assume that members could control their cooperative in an acceptable manner unless they possess knowledge and skill derived at least in part from cooperative education. Again, in order to educate members, communications must have taken place. This reflects member involvement and participation and, in the process, leadership potential must have been encouraged.

As has been said several times, cooperative members have responsibilities that differ from what they would be if they were involved in a proprietary form of business. These responsibilities imply a strong obligation on the part of cooperatives to educate their members. Such efforts include education not only about cooperatives, but also about relevant economic, social, and political conditions.

It was also pointed out that knowledge gained from such education will influence how members vote on cooperative policy questions, their patronage and financial support, and their loyalty to their cooperative when it is challenged by rivals, either in the cooperative or proprietary area, and by those who for any reason lack an understanding of agriculture in general and of cooperatives in particular. Because changes take place in the composition of cooperative membership, employees, and the general public, including politicians in policymaking roles, a continuing education effort is not only important, but necessary.

The importance of education is further emphasized by a cooperative leader who said:

> A well-informed member who understands the organization, its policies, and actions, generally will remain loyal, have fewer complaints, and take greater interest. He will patronize the cooperative when given a choice, stay with the organization when the going is rough, and offer constructive criticism and suggestions. He will inform his neighbors about the organization in terms they understand, serve as an effective salesperson for the organization, help promote new products and services, and be easier to do business with. He will meet his obligations and pay his bills to the

cooperative. An educated member will help stop rumors, defend the cooperative, and develop a favorable climate of understanding between members, employees, and directors. He will promote a progressive attitude and build member confidence in the cooperative and its management. A knowledgeable member will develop a pride among members and within the community in the cooperative as a business organization—and will inform the community of the cooperative's contribution to the local economy.

Communicating with members and keeping them informed, as mentioned before, is necessary to cooperative viability. Cooperative efforts to communicate and thus educate also are necessary for directors, employees, management, and the general public.

EDUCATION FOR ALL

Well-informed directors can be a valuable asset to the cooperative. If knowledgeable, they can reflect by word and deed how their cooperative functions, what its objectives are, and its reason for being. Such directors can ask searching and relevant questions in regard to any policy being discussed. A questioning posture based upon knowledge and overall competence increases the likelihood that sound cooperative policies will be formulated and properly evaluated. Overall, this development of directors through education and communications may result in the cooperative being able to meet the objectives which are established for it. Under these circumstances, there is little, if any, possibility of the creation of a vacuum brought on by a lackluster performance of the board of directors and into which the manager may step to run the cooperative.

Such efforts also pay dividends through better informed employees. In many cases, they are in direct contact with the cooperative's members and with the general public. Well-informed and knowledgeable employees, with knowledge based upon sound information and facts, can be most helpful in creating a favorable image for their cooperative in particular and for cooperatives in general among the various segments of the public with which the cooperative should be concerned.

WHAT HAPPENS
IF COOPERATIVES DON'T EDUCATE?

Some indication of what the situation would be if no formal communications or education programs aimed at encouraging knowledgeable member involvement are carried on by a cooperative is given in a study of member participation made by researchers at the University of Arkansas.

In that study, the cooperative managers were asked their opinions of the most serious problem(s) that would arise if the cooperative failed to maintain a high level of member relations over time. More than half of the 23 managers who responded said that loss of membership would occur. A similar answer was given by two other managers when they said they would expect to lose product volume, in this case, milk. Eight managers cited other problems. The major problems, along with the other problems cited that the managers would anticipate from failure to maintain a high level of education, communications, or member relations, are shown in Table 16.1.

The importance attributed by the managers to member relations and involvement in their cooperatives was further evidenced by the managers' assessments of the greatest strengths and weaknesses of their cooperatives. The two strengths most frequently cited were "hired management and director awareness and leadership" and "member loyalty and support" (see Table 16.2). Two of the three general managers said

TABLE 16.1
Most Serious Problems Resulting from Failure to Maintain a High Level of Member Relations

	Managers listing the problem as the most serious	
Potential problem	No.	%
Decline in number of members	13	56
Loss of milk volume	2	9
Other[a]	8	35

[a]None of the "other" responses were listed more than once. They are as follows: (1) lack of leadership, (2) failure of members to understand cooperative functions and their importance to the market, (3) member dissatisfaction or disinterest, (4) loss of efficiency and higher operating costs, (5) loss of effectiveness of the cooperative in the marketplace, (6) resistance to new or altered programs, (7) loss of enthusiasm among members and employees, (8) loss of member confidence in management, (9) increased problems of organization fragmentation (i.e., more splinter groups), (10) breakdown of regional cooperatives, (11) diminished financial strength, and (12) decreased dairying.

Source: Calvin Berry, William Dabney, and Donald Voth, "Managers' Perceptions of Member Participation in and Control of Selected Large-Scale Dairy Cooperatives," Station Bulletin 868, Agricultural Experiment Station, University of Arkansas, January 1984.

member loyalty and support was the greatest strength of their cooperative, while the third cited hired management and director awareness and
loyalty as the greatest strength (see Table 16.3). As shown in Table 16.3,
other managers, not at the general manager level, gave a much broader
array of opinions on sources of organization strength, reflecting perhaps
the differences in their range of responsibilities.

It is quite apparent that good member relations is recognized as a
major source of cooperative strength and, obversely, a major contributor
to cooperative weakness if it is absent. Lack of member involvement was
the weakness most frequently cited by the managers, as seen in Table
16.1. Lack of communication was also mentioned frequently. It is difficult to visualize a situation in which there is great member involvement
while at the same time there is a lack of communication by the cooperative with its members. The reverse situation, great effectiveness in member communications and poor member involvement, is also difficult to
accept or visualize. It appears logical, therefore, that they go hand in
hand—that it is not likely that a cooperative would, or could, have one
without the other. And since something must be communi-

TABLE 16.2
Greatest Strengths and Weaknesses of the Cooperatives

Item	Managers indicating the item as their greatest strength or weakness	
	No.	%
Strengths		
Hired management and director awareness and leadership	8	31
Member loyalty and support	7	27
Good marketing program	5	19
Sound financial base	2	8
Other[a]	4	15
Weaknesses		
Lack of member involvement	5	22
Lack of communication within the organization	4	17
Incompetent or nonaggressive employees	3	13
Other[a]	11	48

[a]None of the "other" responses were listed more than once.
*Source: Calvin Berry, William Dabney, and Donald Voth, "Managers'
Perceptions of Member Participation in and Control of Selected Large-
Scale Dairy Cooperatives," Station Bulletin 868, Agricultural Experi-
ment Station, University of Arkansas, January 1984. p. 18.*

TABLE 16.3
Greatest Strengths of the Cooperatives, by Manager Level

	Manager level[a]			
Strength	I	II	III	Total
Hired management and director awareness and loyalty	1	3	4	8
Member loyalty and support	2	2	3	7
Sound financial base	0	0	2	2
Good marketing program	0	1	4	5
Other	0	2	2	4

[a]Level I represents the highest level of management and level III, the lowest. All numerals represent number of managers.

Source: Calvin Berry, William Dabney, and Donald Voth, "Managers' Perceptions of Member Participation in and Control of Selected Large-Scale Dairy Cooperatives," Station Bulletin 868, Agricultural Experiment Station, University of Arkansas, January 1984. p. 18.

cated which is substantive and relevant in nature which encourages involvement, we have the process of education taking place. Quality education, communications, and involvement are all part of the same process, and there is much evidence that the result, member loyalty and understanding, is vital to cooperative health.

THE PROCESS OF EDUCATION AND COMMUNICATING

The preceding section has attempted to establish the importance of member education and the role of two-way communication—bottom up and top down—in serving this important purpose. It has been suggested that education or communications, per se, were not the objective, but the ultimate end which should be served in the meaningful involvement of members in their cooperative. It was even suggested that without such involvement, the cooperative was depriving itself of a factor which is widely held to be necessary for long-term viability of the cooperative.

Education and communications resulting in meaningful member involvement are not viewed as substitutes for poor products and services, poor marketing methods, unqualified board of directors, or poor management. Rather, they are viewed as a means of preventing such circumstances. It is difficult to see a situation in which members are meaningfully involved in their cooperative that inferior performance in any of these areas would be tolerated—at least for long.

WHAT SHOULD BE COMMUNICATED?

If education and communications are so essential to member involvement and if member involvement is so crucial to long-term viability of the cooperative, it is obviously important that cooperative boards of directors and management set about performing these necessary functions. Once an appreciation of their importance has been developed by the board of directors and management, what should be communicated to the members? Are there guidelines that will serve to determine the substantive content of education and communications programs?

Perhaps an overall guideline or criterion to use in determining content of education efforts relates to whether providing the information to the members will enable the cooperative to better achieve its goals and objectives. Since the cooperative belongs to the members and since it supposedly was set up to overcome problems perceived by the members, there are no differences in the cooperative's objectives and those of the members.

Once the requirements behind the use of this guideline or criterion are met, the content or substance of communications efforts becomes clear. If it helps in achieving the cooperative's goals and objectives, then set about doing it.

INFORMATION TO BE COMMUNICATED

The board of directors and manager should have some awareness of the extent of knowledge of the members, based upon past education efforts and perhaps evidenced by the extent of meaningful involvement of the members in the cooperative. Having such knowledge, they should be able to tailor education efforts to fit present needs. Such efforts, based upon long-term objectives and criteria, could result in programmatic efforts designed to be cumulative in impact. This program would start where the last one ended in an effort to achieve objectives in an efficient manner and avoid duplication of efforts in presenting the same or similar subject matter again. Achieving such efficiency, however, may call for segmenting the membership on the basis of how long they have been members and achieve some degree of homogeneity in this manner. It is to be remembered, however, that composition of the membership changes over time, and it is better to err on the side of duplicative efforts than to risk missing a portion of the membership in the education and communications effort.

COOPERATIVE INFORMATION

One area that always falls within the guidelines mentioned relates to information about the cooperative. There should be little, if any, concern about overlapping or duplicating of program efforts because this is one area that should be covered many times. The membership changes and relevant information about the cooperative changes.

Every so often a cooperative should conduct education programs covering areas such as the following:

1. What a cooperative is—its unique nature and its reason for being.
2. Background and history of cooperatives in general and of this cooperative in particular.
3. Objectives, goals, policies, programs, and philosophy of the cooperative.
4. How the cooperative operates—stress the management trio discussed in Chapter 8.
5. How their cooperative is structured.
6. How members can get information about the cooperative, its methods, and any problems they might have.
7. The voting, election, and representation system used by the cooperative.
8. The bylaws of the cooperative.
9. How the cooperative is financed and why it is financed in this way.
10. The role of the board of directors.
11. The role of the manager.
12. How members' equity is redeemed and why it is done in this way.
13. The financial statements—balance sheets and operating statements for relevant periods.
14. Taxes paid by the cooperative.
15. The relative competitive position of the cooperative vis-à-vis other businesses.
16. How are product prices determined?
17. What is pooling?
18. What happens to net savings or margins?

19. What general governmental policies affect the cooperative now and what is expected?
20. How does the consumer fit into all of this? Is our only concern tied in with production and letting our cooperative do the marketing?
21. What's the difference between marketing and selling?
22. Information about any current problems being faced by the cooperative.
23. Measures of performance used by the cooperative and what they indicate for the past few years.
24. Trends in sales, share of market, and other areas.

It was mentioned that no programmatic effort such as communications can be used to cover up a continuing poor performance of the cooperative. Fundamental to any effective cooperative communications program is the degree of success the cooperative enjoys in meeting the objectives established for it. Nothing can make poor management, inadequate performance, or unexplainable bad years palatable to anyone who has an interest in the cooperative's well-being such as docs the member. A high level of operating performance is absolutely essential, and knowledgeable member involvement usually directly associated with effective educational and communications efforts serves a very complementary role in achieving the desired performance.

HOW DOES A COOPERATIVE COMMUNICATE AND EDUCATE?

We've stressed the extremely important role of education and communications in contributing to the well-being and smooth functioning of the cooperative. Our next question relates to how effective communications are carried out. What methods are used?

Many different ways can be used in communicating with people. These include personal contact, written messages, and electronic methods such as radio and television.

Personal contact is generally considered the most effective means of cooperative communication. This type includes one-on-one personal contact, various kinds of group meetings, open houses and tours of facilities, member committee meetings, and meetings of the members with the manager, board of directors, and employees. The strength of this type of communication ties in with the fact that there is usually an opportunity, if well-planned, for feedback and reaction on the spot. As the group gets larger, of course, such opportunities for feedback grow more limited.

Other types of communications methods such as newsletters, direct mail, member magazines, newspapers, annual reports, and personal letters are used. Television and radio are also being used extensively.

As cooperatives grow larger and the lines of communications between the member, the board of directors, and management grow longer and more difficult, the ingenuity of the management and the board of directors is challenged. There is the psychological need on the part of the members to feel that their voices are being heard. Perhaps there is no expectation that each and every suggestion or comment made by the member is acted upon, but somehow there has to be the feeling that if I wish to say something, I will be heard.

In addition to satisfying this perfectly natural human desire to feel that one is being heard, there is the necessity of having knowledgeable member understanding and involvement in the affairs of the cooperative. This unique feature of member involvement as a part of the management trio has great strength if its potential is exploited. The instant it is viewed as a cumbersome, awkward feature and it is short-circuited in any manner, the cooperative is treading on dangerous ground. This is an essential and distinctive feature of cooperatives and it must be respected.

WHO COMMUNICATES?

In addition to those who have specific responsibilities in education, communications, and member relations, anyone who is connected with a cooperative in any way is a communicator. This is true of the employees, the receptionist, the telephone operator, the milk hauler, the field person, the tank truck driver, or anyone who because they are associated with the cooperative and come into contact with people become the cooperative in the eyes of the member. In a recent study of a leading cooperative, the milk hauler was indicated as being a major source of information. This means, of course, that each and every person and group associated in any way with the cooperative should be aware of their implicit role as a communicator. Depending upon their appreciation of the importance of this role and how well they are equipped with adequate and correct information about the cooperative, they serve useful roles.

COMMUNICATE EVERYTHING?

The overall general policy of a cooperative is that the members are owners of the cooperative and thus have a right to be kept fully informed about all issues and questions. There are situations, however, in which decisions have to be made in this regard. The question is whether the members' need for and right to have current information, if strictly adhered to, may not be in the current best interest of the cooperative and thus not in the best interest of the members. Again, it is well for management and the board of directors to establish guidelines or rules that will be followed in making such decisions. The members, of course, will be advised completely about the guidelines along with hypothetical examples of circumstances when information would be withheld, at least temporarily.

The overriding criterion that should provide the basis for the guidelines is whether the members not immediately having the information will result in decisions by the board of directors and management which will not be in the best interest of the cooperatives. Or, put another way, will the members having complete current information about a situation be helpful to the cooperatives and not be helpful to the competitors of the cooperatives?

POLICY AND PROGRAMS
IN COMMUNICATIONS

This brings us back to our consideration of the distinction we made previously between a policy and a program. We pointed out that a policy was a basic overall objective of the cooperative usually stated in a general way. A program was a plan of action to carry out the policy. Members must be kept completely informed about cooperative policy in all areas. They would not expect to be kept currently informed in regard to detailed changes in programs designed to carry out the policy.

For example, it may be a policy of the cooperative to serve the fluid milk requirements of the milk handlers in a market and balance the marketwide supply and demand requirements. The members should be completely informed about this policy, its rationalization, and its realization.

However, in a situation in which the cooperative finds it necessary to seek additional markets in order to satisfy the supply–demand balancing policy, informing the members on a day-to-day basis would serve no reasonable purpose.

As a general rule, the more controversial the issue the greater the need for the cooperative to provide sound and reliable information regarding the issue to its members. Nothing is potentially more destructive to the cooperative than to leave the issue open to rumor and gossip. Nothing can be more supportive and conducive to great member loyalty than the provision of frank, open, factual information by the cooperative to the members. If the situation can be anticipated and information can be provided on a reasonably sound basis, before the issue becomes public, this would be in order.

Member loyalty, understanding, and meaningful involvement in their cooperative is directly associated with the effectiveness of the cooperative's education and communications efforts. Effective efforts in these areas are no substitute for inefficient operations or poor performance, but assuming the idea and formation of the cooperative are soundly based, the two go hand in hand.

OUTSIDE THE COOPERATIVE

There are, in addition to the members, employees, and management staff personnel, other segments of the public with which cooperatives should be concerned from an informational standpoint. These include potential members and the general public. This may also very well include young cooperative members or the family of a cooperative member who, for some reason, may not have been involved in the affairs of the cooperative. They may have taken the cooperative for granted, since it had always been there, and may have been content with letting the parents or older members of the family be involved. As indicated at the end of this chapter, this group represents the leadership foundation upon which the cooperatives will have to rely in the years to come. For this reason, special informational and communications efforts are needed with this group despite the fact that it is not a part of the general public. It goes without saying that the general public merits continuous, special efforts on the part of the cooperative in bringing about an understanding and appreciation of this institutional arrangement. This is an area that has not been handled in an adequate manner as evidenced by the public's lack of understanding of agriculture in general and farm cooperatives in particular.

THE GENERAL PUBLIC

As implied in Chapter 1 and passim, there probably has never been a time when there was less understanding of agriculture in general and of cooperatives in particular than at the present. This is reflected by some of the statements frequently made by the public regarding policy and programs being considered for agriculture and by some who consider cooperatives as not being a legitimate part of the capitalistic economy.

This phenomenon is a natural outgrowth of the shrinking agricultural population as a percentage of the total population. Fewer and fewer people have had an association with agriculture, with rural areas, and with agricultural cooperatives. It therefore behooves those who understand agriculture and cooperatives to provide sound and factual information regarding the economic bases for cooperatives and the role they can play in meeting the food and fiber needs of society. This task can be done in other than a self-serving way. Such efforts can bring about a greater understanding and will serve the best interests of all.

There are many segments of the so-called general public with whom the cooperative should communicate. These should be viewed as special targets, and special efforts should be mounted to serve them. They include legislators, agencies of government at various levels, the news media, schools at all levels from elementary through college, churches, and civic clubs.

These groups have some degree of influence on others, including cooperatives. Perhaps because of lack of information or because of having incorrect information, their influence can be negative. Many groups, such as legislators who are charged with establishing agricultural policy and cooperative policy, are especially in need of factual information which can help them in formulating policy and programs in the public interest. Without an understanding of what is involved, it is too much to hope that soundly based policy will emerge. An advocacy role is not suggested, but communication and education efforts on the part of cooperatives are strongly suggested.

POLITICAL ACTION COMMITTEES

An interesting phenomenon in the cooperative arena has arisen over the past few years. Political Action Committees, known as PACs, have been formed to inform legislators and public office seekers about agriculture and cooperatives. Positions of the candidates are sought in regard to various issues, and these are made known to members in order to help them in deciding how to vote at election time. Assessments of members at various rates have provided the funds for this type of activity, but, in most cases, such assessments have been on a voluntary basis and not mandatory.

It appears quite obvious that the cooperative and agricultural community does not feel completely comfortable in carrying out this type of activity. They apparently have few, if any, qualms regarding such activities by labor, education, or the medical profession, but are not yet willing to openly endorse and engage in such efforts themselves.

Perhaps an approach that emphasizes the educational and communicational aspects of this activity rather than the political could overcome the reluctance of agriculture to go all out in what is now viewed as the political arena. By providing basic information regarding agriculture and cooperatives, the structure of agriculture, and the economic raison d'être of agricultural cooperatives, those who seek and are elected to public office and are faced with the extremely important task of formulating policy in these areas might be better able to perform the task in such a way that the interests of everyone, the cooperative and the general public, can be served. At least the dulling and dampening effects of lack of inadequate, or incorrect information might be partially ameliorated.

YOUNG COOPERATIVE MEMBERS

Mention has been made of the importance of educational and communications efforts being directed at various groups. One of these is the younger cooperative members who have in many cases never been really involved in the affairs of the cooperative. Seemingly, they have been content to let their parents be involved, and they themselves stay in the background. Of perhaps greater potential damage and loss of future potential leadership capacity, the parents have been content to let their children play passive roles.

There is evidence to indicate that younger cooperative members' attitudes and perceptions regarding the cooperative are different from those of the older members. In many cases, they may feel that the cooperative has not offered or provided a chance for them to participate in its affairs and work up to positions of leadership. It is very unfortunate if this indeed is the case. Let's now examine the results of a special effort made by a cooperative to encourage involvement of potential young leaders.

YOUNG LEADERSHIP—
A SPECIAL EFFORT BY ONE COOPERATIVE

Regardless of attitudes or positions, there is some evidence of a growing appreciation of the fact that the future of the agricultural cooperative is in the hands of today's young potential leaders. The question is not whether our cooperative institutional arrangement will be within their hands, but how well they will be equipped to handle it when they must take over. There is also some evidence that current leaders, boards of directors, and management are recognizing more strongly that they have an obligation and responsibility to help potential leaders by equipping them to carry out their future roles. Further, they are taking steps to meet this obligation and responsibility.

In order to help in meeting this obligation, the board of directors and management of a cooperative arranged to conduct meetings of young leaders in appropriate geographic areas. Such an activity was suggested by members at local and district meetings and was offered in the form of a resolution by delegates in their annual meeting.

An invitation was sent to potential leaders within relevant age groups. It was stressed that this was a special meeting for special people and that the cooperative recognized clearly that its future and the future of the cooperative movement in general was in their hands.

Some degree of structure was built into the meetings by including a few relevant topics such as why we have cooperatives and their history and economic basis. The overall format, however, was designed to encourage informality and meaningful discussion. The young leaders were encouraged to be very frank and candid, to raise questions, to enter into the discussions, and to raise any points or questions they wished. A member of the board of directors of the cooperative from the geographic area where a meeting was being held was present to respond to questions raised by the young leaders. The meetings were well attended and lasted most of one day. An attempt was made to evaluate the effectiveness of the meetings as measured against the stated objectives. The program was developed with the help of some of the young leaders in a session arranged for the purpose of getting their input into what they thought was important as a part of the program and in regard to the objectives and goals of such an effort.

EVALUATION

A major area of importance in attempting to formally evaluate the effort made by the cooperative to face up to its obligation to be concerned about future cooperative leadership relates to whether the potential young leaders are receptive to such efforts. Are they interested in preparing themselves for leadership roles? Are they concerned about future cooperative leadership and their possible role in contributing in this area? Are they interested in cooperatives? Do they understand why and how their cooperative came into existence? If they indicate that they have no interest or no concern and are not receptive to educational efforts on the part of their cooperative, this is highly significant and suggests certain assessments and directional movements. If they are receptive, this is also significant and provides direction for future efforts.

As shown in Table 16.4, about two-thirds of the young leaders were very positive in their evaluation of the meetings. Significant percentages found them informative and educational. There were no negative responses.

This seems to suggest a high degree of receptivity on the part of the potential leaders to participating in such meetings and attempting to equip themselves for performing in leadership roles. As will be stated again later, this represents a major challenge to our present cooperative leaders.

Another question used in evaluating the effectiveness of the meetings was, "What do you see as your role in your cooperative over the next 5 years?" Responses to the question included the following: (1) be in-

TABLE 16.4
Response to Question, "What Are Your Reactions to Today's Meeting?" (Meetings, Young Cooperators, January 1983, $N = 544$)

Category	Number of respondents	Percentage of respondents	Percentage of responses
Informative	224	41.2	21.6
Interesting	74	13.6	7.1
Educational	162	29.8	15.6
Rewarding	8	1.5	0.8
Excellent	12	2.2	1.2
Very positive	361	66.3	34.6
Beneficial	28	5.1	2.7
Impressed	11	2.0	1.1
Enjoyable	58	10.7	5.6
Do it again	101	18.6	9.7
Totals	1039[a]	191.0[a]	100.0

[a]More than one response given by respondents.
Source: Young Cooperator Meetings, January 1983.

formed, (2) be involved, (3) participate, (4) more commitment, (5) be supportive, (6) serve my cooperative, (7) understand cooperatives, and (8) encourage leadership.

By far the highest percentage response was that their role in their cooperative over the next 5 years was to be involved. The next highest response category was to be informed.

Another question was, "How can all of us as young cooperators working together carry out the roles which were indicated in responding to the first question?" This question was designed to push them a bit deeper into thinking how they could, in fact, carry out the roles they saw for themselves.

As shown in Table 16.5, the highest percentage response relates to participation in the activities of the cooperative in the process of supporting it in a knowledgeable manner.

High response areas had to do with keeping informed and communications. We have to be informed, we have to communicate, and we have to talk were positions suggested many times in the evaluation process and in the discussions at the meetings. Understanding of cooperatives, their basic foundations, and why we have them was recognized as critical in making it possible for them to fulfill their roles as they saw them.

It was obvious that the young people felt very strongly that they should be informed, that they should talk and communicate, and that

TABLE 16.5
Response to Question, "What Do You See as Your Role in Your Cooperative Over the Next Five Years?" (Meetings, Young Cooperators, January 1983, $N = 461$)

Category	Number of respondents	Percentage of respondents	Percentage of responses
Be informed	218	47.3	24.1
Be involved—participate	370	80.2	40.8
More commitment	37	8.0	4.1
Be supportive	36	7.8	4.0
Promote diary products and my cooperative	102	22.1	11.3
Serve my cooperative	38	8.2	4.2
Understand cooperatives	19	4.1	2.1
Encourage leadership	31	6.7	3.4
Communication	54	11.7	6.0
Totals	905[a]	196.3[a]	100.0

[a]More than one respondent and response.

Source: Young Cooperator Meetings, January 1983.

they should raise questions and understand the why of their cooperative and how it worked. Lines of communications should be kept open so they could keep themselves informed and make intelligent inputs into the workings of their cooperative. This, they felt, would make it more nearly possible for them to play a constructive role over the next 5 years.

MEANING AND IMPLICATIONS

One of the comments or reflection of feelings which came forth at all the meetings was that the young potential leaders had taken this institutional arrangement for granted. They pointed out that they had never been involved in starting a cooperative. They had pretty much sat on the sidelines and let their parents be the involved ones so far as the cooperative was concerned.

They suggested that it had been driven home to them for the first time that it was important for them to understand how cooperatives came about and why we have them. It is necessary, they concluded, that as members they should be involved and understand as well as appreciate the workings of their cooperative and what is necessary for it to work properly. They pointed out that they should understand that the cooperative institutional arrangement is unique and is a special form of business arrangement in our economy. It is designed to serve the agricultural sector of our economy which is in itself unique. One young leader couple suggested that there is reason to believe that young potential leaders are anxious and are grasping for ways and means of being meaningfully involved in their cooperative.

To the extent that these reflections are true, they represent a major challenge to those who are presently in leadership roles in cooperatives. To provide the information, the incentive, the format, the scene, and the setting in such a way that this craving and desire to be instrumental in causing this type of institutional arrangement to carry on in its most meaningful sense is a major challenge. It appears that cooperative leaders should give this top priority on the cooperative agenda.

Young leadership development, of course, is only one facet, albeit one of the most important, of the objectives involved when resources are committed to education of members. There are many segments of the within house part of the cooperative. The board of directors, the employees, and the membership at large all reflect the cooperative's image and have an impact on its ability to perform in such a way that its objectives will be achieved. The importance of meaningful member involvement in the affairs of the cooperative cannot be overemphasized.

Purposeful efforts in communications and education can bring about this involvement.

REFERENCES

Barton, D. G. 1984. The University Mission and Cooperative Education. NICE, Bozeman, MT.

Isaksen, G. P. 1983. Political Support for Farmer Cooperatives. WP-68. University of Wisconsin, Madison.

Leith, W. G. 1980. Reflections on Leadership in Cooperatives. NICE, Pennsylvania State University, University Park.

McBride, G., and Burnett, J. 1983. Encouraging and Developing Agricultural Cooperative Leadership, A Case Example. Staff Paper No. 83-24. Department of Agricultural Economics, Michigan State University, East Lansing.

Rust, I. R. 1966. Effective Information Devices for Cooperatives. FCS Educational Circular *29*.

U.S. Department of Agriculture, FCS. 1971. Recruiting, Training, and Developing Workers for Farmer Cooperatives. FCS Information *77*.

U.S. Department of Agriculture, FCS. 1972. Opportunities in the Co-op Business World: A Leader's Program for Youth. FCS Information *80*.

U.S. Department of Agriculture, FCS. 1976. Advising People about Cooperatives. FCS, *PA-1147*.

U.S. Department of Agriculture, FCS. 1983. Cooperative Education and Training. Cooperative Information Report *1*, Section 10.

TO HELP IN LEARNING

1. Conduct a survey among various groups with which you come into contact and determine how "educated" they are in regard to significant aspects of agricultural cooperatives.

2. Find the names of two or three young members of cooperative families and discuss with them the extent of their involvement in their cooperative and their knowledge about it.

3. The director of the U.S. Office of Management and Budget recently made some rather negative statements in regard to past U.S. agricultural policy and programs. Determine what the gist of his state-

ment was and comment in a constructively critical and professional manner.

4. Quiz your peers in regard to farm organizations to which their parents or other family members may belong. Determine what educational programs are conducted by the organizations.

5. On the basis of your own observations, what efforts are being made by farm groups through their cooperatives to provide relevant educational information about agriculture and agricultural cooperatives?

DIRECT QUESTIONS

1. In your judgment, should member education be a cooperative principle? Explain.

2. What is a proxy? What are examples of proxies used in this chapter?

3. If we assume that cooperative directors have been selected and elected, as was suggested earlier, there is no need for director education. Comment.

4. Why should cooperative employees be knowledgeable about cooperatives?

5. What are some of the indicators of effective cooperative efforts in cooperative education?

6. What are some of the indicators of lack of effective educational efforts?

7. What is meant by bottom-up, top-down lines of communications?

8. What is the guiding principle in determining if some information should be at least temporarily withheld from members?

9. Give an example of information that might be temporarily withheld from members.

10. What guideline should be used in deciding what to communicate to members?

11. How might duplicative educational efforts be avoided?

12. Are educational efforts a substitute for poor performance of the cooperative?

13. Is current successful performance of a cooperative a substitute for poor or no educational effort? Explain.

14. State your policy as tied in with your answers to questions (12) and (13).

15. Most cooperatives have education divisions or sections. Is this the only part of the cooperative that communicates?

16. Again, distinguish between a policy and a program.

17. If an issue is very controversial, the cooperative should keep quiet and let it go away. Comment.

18. Why do some potential young cooperative leaders feel apathetic about their cooperative?

19. Many national media commentators are negative about agriculture. What does this indicate about agricultural communications and educational efforts?

20. Discuss PACs. Should they exist? If yes, how should they be paid for? If yes, what should they do? Explain fully.

TYING-TOGETHER QUESTIONS

1. Is there anything in the framework within which we've been considering agricultural cooperatives which speaks to the question of the importance of education, communications, and leadership development?

2. You are the director of information and communications for a cooperative. Do the following:
   a. Develop your basic overall objectives as the person responsible for communications, education, and leadership development—the policy as established by the board and manager.
   b. Plan a 5-year program to carry out your objectives—the programs in outline form.
   c. Detail your first-year effort in terms of content, audience, geography, actors, and other areas.
   d. Detail evaluation of effort—procedures used and use of results.
   e. Indicate how succeeding years of your 5-year plan might be changed and for what reasons.

3. Your congressperson is considering how to vote on a bill to change Section 2 of the Capper–Volstead Act and asks you for help in understanding what's involved. Set up, in detail, an educational program for your legislator.

17

Cooperative Performance—
Its Goals and Measurement

It would be interesting and perhaps worthwhile for us to stop at this point and give some thought to how, if possible, cooperatives can assure success of their operations. Perhaps we might prepare a paper on the topic, "The Profile of a Successful Cooperative." We've set forth in detail the steps in forming a cooperative after it was determined that there was a need for a cooperative and that it was economically feasible. We then covered areas relating to management, financing, and other aspects in the how section of the book.

But once all these steps have been taken, how do we really know if our cooperative is performing well? What constitutes success and is success the same for all cooperatives? Let's now use the knowledge we have gained in all these areas and focus on the area of performance of our cooperative. Assume we have our cooperative going, and focus for awhile on the question, "How are we doing?" Are we providing the services we felt were needed when we were considering whether a cooperative should be started? Are we performing them well? Could they be performed better?

PERFORMANCE—CRITERIA

Before we can proceed very far in our concern with or questions regarding how we're doing, it is necessary to define or redefine our goals or objectives. Why did we start the cooperative in the first place and have our goals and objectives changed to any extent since that time?

Once we know our reason for being, our goals and objectives, we want to measure how well we are achieving them. Thus, we want to measure our performance. Before we measure performance, we have to define what it means and then figure out some way to measure it. Once this is done, we should concern ourselves with full use of the findings in changing our direction if our performance is found to be less than satisfactory. While all of this is no easy task, it is an essential step in finding our strengths and weaknesses and gaining insights into how the performance of our cooperative may be improved.

OUR ANALYTICAL FRAMEWORK

In order to guide our efforts in our attempt to measure the performance of our cooperative, let's adapt a portion of the model of industrial organization which we used in Chapter 1. Reference to that model shows that the two bottom sections are labeled conduct and performance. Let's use those two sections of the model, along with an evaluation center section, as an aid in the process of measuring our performance. The model framework is shown in Fig. 17.1.

ELEMENTS OF THE FRAMEWORK

The same line of causation can be assumed in this revised model as was assumed in the original model, that is, certain areas of conduct will lead to certain kinds or degrees of performance. The most important task and by far the most difficult, however, is to find elements which logic would suggest fit appropriately into the designated areas. What elements have an influence on whether the cooperative's performance is as is desired, and also, what is performance itself? Once we have the appropriate elements and we know what we're trying to measure, we may have come up with a framework that will permit us to say something about the performance of cooperatives from the observation and measurement of the conduct elements. So let's fill in the boxes of our framework.

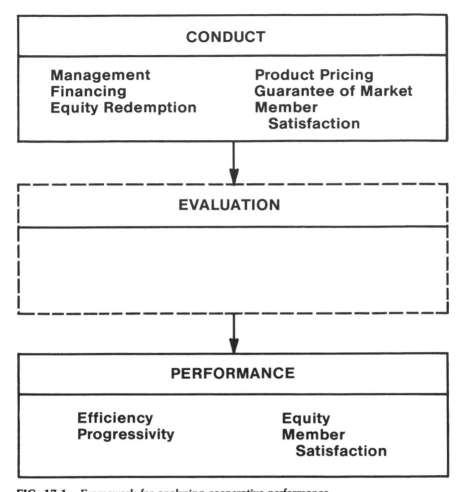

FIG. 17.1.. Framework for analyzing cooperative performance.

Source: Adapted from Scherer, F. M. 1980. Industrial Market Structure and Economic Perform-ance, 2nd Edition, p. 4. Copyright by Houghton-Mifflin Company, Boston. (Used by permis-sion.)

PERFORMANCE

Let's take the bottom section of our framework first. This is a logical step, since performance represents goals or objectives or ends toward which we should be striving in operating our cooperative. Of course, the cooperative was, very simply, set up in the beginning to provide a service or services that could not be made available at all or as efficiently by individual producers. Providing these services is thus an objective, but it stands to reason that we should know whether they are, in fact, being provided. Our concern here is with determining how well they are being provided. We need a proxy for this determination.

Perhaps we could use efficiency as a proxy for measuring our performance in the area of providing the service(s) that prompted the starting of the cooperative.

Another ultimate objective or end relates to whether we are alert to possibilities of better ways of providing the service(s). Are we innovative or are we satisfied to use the same techniques even though better ones are available or could be made available? Do we encourage new ideas? Do we encourage experimentation and innovativeness? Perhaps we could call this progressivity and then seek proxies to use in measuring it.

Equity can very well be considered as an overall objective of the cooperative despite the fact that it is a slippery concept and is not easily measured. Proxies for its measurement are not readily available. It has a fairness connotation and relates to the distribution of rights, responsibilities, returns and costs, and the sharing of benefits in proportion to costs borne. The concept may even be permitted to cover external relationships between the cooperative and the public sector. This relationship, of course, conjures up questions about the cooperative's responsibilities in keeping the public equipped with adequate and reliable information regarding its view of the public interest and its efforts to bring about a high degree of conformity between its own interests and those of the public. This is done in a true public relations framework and is not designed to be self-serving, although it is likely that it will be helpful to the cooperative.

Finally, an overall objective of the cooperative corporation relates to its ability to contribute to the economic well-being of its members. Presumably, this means the cooperative is providing the service or services that those who started the cooperative thought were needed. This could be designated member satisfaction in our analytical framework.

FOUR STANDARDS SUGGESTED

We are suggesting strongly that cooperatives should constantly be in an evaluative mode and be concerned at all times with questions relating to how they are performing. Before performance can be measured, it is necessary to define it and then measure the cooperative's operation against overall standards or objectives which seem appropriate. We have suggested four such standards—efficiency, both allocative and pricing, progressivity, equity, and member satisfaction—as perhaps being appropriate for use as areas to evaluate in assessing the performance of our cooperative.

This does not mean that other areas might not be used as appropriate ones or that additional ones may not be added. The thrust here is that cooperatives should concern themselves in a systematic manner with some type of evaluation or measurement process, and in order to do this, appropriate standards, objectives, or ends must be established against which the cooperative's performance can be measured. So we place efficiency, progressivity, equity, and member satisfaction in the bottom part of our framework and then attempt to find actions or conduct which we think will impact either positively or negatively upon them. Then, of course, we'll have to measure in some way the extent of the impact.

CONDUCT

Implicit in the previous discussion is the suggestion that the performance of a cooperative relates to the consequences flowing from conduct on the part of the cooperative, both internally and in its reaction to its external environment. Evaluation of performance, as stated, involves the appraisal of the extent to which the decisions in the conduct area on the part of the cooperative's decision makers stimulate results that are consistent with stated criteria, which we are calling performance dimensions. We have suggested that the performance dimensions should be stated in rather broad terms, as is the usual case in any discussion of means–end relationships. We have also suggested that the dimensions themselves usually do not lend themselves to direct measurement and that in most cases, proxies must be found which can be measured and which, when measured, will provide some indication of the extent to which the performance dimensions are being achieved. But our next task is to find or indicate possible areas of conduct on the part of the cooperative which are aimed at achieving the stated performance objectives.

Our concern here is with listing appropriate areas of conduct in which the cooperative can engage in trying to meet its objectives, objectives which we have wrapped up into four bundles—efficiency, progressivity, equity, and member satisfaction. What conduct options are open to cooperative management in trying to achieve these goals?

Rather than listing the myriad of things a cooperative can do in the area of conduct, let's list the areas within which the actions taken might most appropriately be placed. For example, one of the areas we'll place into the conduct option box is management. Efficiency and member satisfaction may be performance goals most directly impacted by management practices.

Again, referring back to the areas we discussed as being uniquely cooperative, we can use financing as the area for another box in our conduct options. This also impacts upon efficiency and member satisfaction performance criteria.

Equity redemption policy and programs might be used in our box of conduct options as a proxy for the performance criterion we have labeled equity.

Our last box of conduct options, which we've labeled member satisfaction, is not completely separated from the others, as is also true of all boxes. However, product pricing, guarantee of market, and so on would fall into this category. Since so many types of conduct impact on this area, let's call it member relations.

We're ready now to place the designations or areas we've chosen as appropriate into their segment of our analytical framework. Let's place them into the framework shown as Fig. 17.1. Of course, the middle box, evaluation, will remain a question mark, since it is suggestive of our procedure rather than of more tangible elements.

BOXES NOT DISCRETE

It should be pointed out that it is recognized that each of these boxes is not separate and distinct from all the others. There's no intention to suggest that conduct engaged in, for example, shown in the management box, has an impact on only one criterion listed in the performance standards. As a matter of fact, it may well be that only one box, management, is needed. Also, all conduct options could easily be included in this box, and actions taken under this heading would impact on all the performance criteria.

In the interest of streamlining the process a bit and in emphasizing the importance of constant evaluation of performance, it was thought that breaking the areas into parts might be useful.

EVALUATION

We now have our conduct and performance boxes filled, as shown in Fig. 17.1. It is noted in the framework that a section using dotted lines is shown between the conduct and performance boxes. A dotted line also connects the two boxes.

The middle box is designated as evaluation and the dotted lines indicate that it is not a definitive area as are the conduct and performance areas. However, it may be just as important as the others, since it is only by systematically and professionally engaging in the evaluation procedures that effectiveness of the conduct options in achieving the performance objectives can be assessed. The dotted line on the right side of the diagram which connects the conduct and performance boxes recognizes the fact that the flow of causation may not be a one-way street. Usually, we think of the flow as being from conduct to performance, but there may well be impacts flowing in both directions.

In the same vein, it should be pointed out that there is no intention to suggest that the boxes in the two areas are discrete, horizontally. Management practices certainly have an impact on member satisfaction and a conduct option taken in management may well impact on all the performance criteria. The intent, however, as stated, is to drive home the point that keeping in mind certain performance standards is essential and that conduct options exercised will impact upon them. Despite the implication and suggestion that the dotted line box indicates an area of less importance than the others, this may not be true. It is only through evaluation that we can know in some definitive way whether the conduct options being used are in fact the ones which should be used. There's a sensing or feeling of what is being accomplished, of course, but without systematic evaluation of action impact on stated performance standards, we don't really know how we're doing.

It is also possible that the area of evaluation may be as difficult, or even more so, to handle satisfactorily than the others. It is remembered that we purposely defined the performance standards in rather broad, intangible terms. Efficiency, for example, means a number of things to people. Equity, we said, is one of the most slippery concepts that comes up.

This means that we can't and won't even try to measure directly the degree to which the cooperative has met the standards of performance which are designated as appropriate. They're of such a nature that they can't be measured directly and that means only indirect assessments can be made. This means that we'll have to search for proxies for the standards which, when used properly, will give a reliable approximation of the degree to which the performance standards have been met. We'll now turn to the task of seeking such proxies.

PROXIES FOR MEASURING EFFICIENCY

First, we must define efficiency. As usual, there are several types of efficiencies and each one has a different definition.

Efficiency in the engineering sense involves input–output ratios. It implies that the concern is with an output of goods and services being as large as possible in relation to the resource inputs. The greater the output, other things being equal, in relationship to the resource inputs, the more efficient the operation. The use of appropriate technology, the appropriate plant size(s), and of adequately trained and appropriately skilled human resources are implied in this concept of efficiency.

There is also the concept of pricing or allocative efficiency. This refers to the extent to which prices accurately reflect demand preference and long-run average costs. In addition, the use of contracts and government programs may be reflected.

If we wish to go further and get into areas more difficult to measure, in that proxies are more difficult to find, we could mention social efficiency.

This is a kind of "good citizen" concept and relates to areas that are not captured in the usual balance sheet or income statement calculations. Such external factors as environment, pollution, and soil erosion are involved here. Despite the fact that proxies for these areas are difficult to come by and we probably won't include suggested proxies in our compilation of conduct options, the good citizen image is invaluable from a public relations and public interest standpoint. Let's now seek proxies which we might use to measure engineering efficiency. These include the following:

1. Input–output studies—human and material
2. Budgets
3. Financial analysis
4. Liquidity—current assets divided by current liabilities and liquid assets divided by current liabilities
5. Inventory turnover
6. Number of days sales in accounts receivable
7. Salaries and wages—number of employees
8. Gross margins
9. Net margin
10. Ratio of net income to total assets
11. Ratio of net income to member equity
12. Ratio of total liability to member equity
13. Long-term liabilities to member equity
14. Fixed assets to member equity
15. Volume of products received and trends in receipts
16. Marketings, volume, trends
17. Procurement—supplies, products, inputs

18. Personnel—number, trends, training
19. Borrowings—internal, external
20. Products bought per member and sales per member
21. Daily sales and product procurement volume
22. Number of patrons—member and nonmember
23. Average products supplied or sales per member–customers
24. New customers—patrons
25. Hundredweights of product handled
26. Accounts receivable
27. Aging of accounts receivable
28. Accounts payable
29. Amount of past due accounts recovered
30. Number of employees—new—turnover
31. Number of members
32. Delivery per member
33. Total business audit
    a. Management
    b. Board of directors
    c. Organizational arrangement
    d. Employees
    e. Physical facilities
    f. Inventory
    g. Marketing and procurement practices
    h. Transportation
    i. Financial
    j. Plans and budgets
    k. Office procedures
    l. Credit policy

It is easy to see that the engineering concept of efficiency lends itself to measurement by use of many proxies. There is no suggestion that all the proxies be used in an evaluation at a particular time. Many of those suggested above have a current or very short-term performance connotation. The various ratio analyses suggested fall into this category. They are appropriate for particular uses and should be selected and used with a particular purpose in mind. Perhaps all of these would fall into the management box of conduct options. They would also be likely to impact upon other performance criteria in addition to efficiency.

Others are focused on a longer-term orientation, and this is absolutely essential. Feasibility studies can foretell the usefulness of new buildings or other forms of business expansion by the cooperative. The role of long-term planning and its essential nature will be covered later. It is

emphasized, however, that the reason behind evaluating the cooperative's efficiency is not efficiency, per se, but only as efficiency contributes
to the degree to which the cooperative's performance is in accord with
the goals and objectives established for it. Whether proxy measures are
current or long-term oriented, the reason for their proper use as diagnostic and prescriptive tools is the same.

PRICING EFFICIENCY

As previously stated, this concept of efficiency involves pricing or
allocative efficiency. It might even be expanded to include exchange
efficiency. They have to do with whether prices paid for products and
charged for sales reflect demand and supply conditions and are close to
long-run average costs. They include consideration of transaction costs
in the process of exchange. Market coordination is involved.

It is quite likely that purposeful examination of many of the proxies
for measuring engineering efficiency mentioned previously will shed
light on this concept of efficiency. In addition, measures of price levels
and trends would serve our purposes here. Price stability and reasons
bringing about any instability which might be found as measured by
various price series and trends would be helpful in our diagnostic en-
deavor.

SOCIAL EFFICIENCY

Awareness of and concern with some of the factors that are not
ordinarily captured on the balance sheet and income statements of
business organizations is a relatively new phenomenon. Few, if any,
proxies for specific measurement of the cooperative's performance in
this area are readily available. Perhaps it is sufficient if boards of directors and management of cooperatives remain aware of today's concerns
with our environment and its use. To the extent that the cooperative or
any other form of business has the potential for contributing to the
lessening of the quality of our environment and waste of our resources,
and exercises that potential, it may not be shown on its current balance
sheet, but it will be a long-term liability.

PROGRESSIVITY

Our next overall dimension against which performance of our cooperative will be measured is progressivity. This suggests a concern with

whether ways and means are available to the cooperative to enable it to meet its objectives more completely and economically from a quantity and quality standpoint than it is now using. Is there technology available which would be feasible for the cooperative to use? In short, are there better ways to serve our members and more completely achieve our goals?

It is assumed that the position would be taken by the board of directors and management that the operations of the cooperative should be progressive, that advantage would be taken of opportunities opened up by science and technology for increasing output per unit of input and making available to members the services they want as efficiently as is economically feasible. This is assumed, but our question here is how we know if that is being done. Again, we seek proxies we can use in measuring this dimension of performance.

First, we have to be constantly abreast of what the needs of our members are in terms of quantity, quality, timeliness of service, and so on. This suggests constant interaction and communication with them. Many of the short-term measurements we suggested as proxies for engineering efficiency may provide insights into areas where innovative changes in service(s) and/or their delivery are needed. Suggestions for better ways of doing things and for better things to do should be encouraged on the part of employees, members, and others who might be in position to make suggestions.

This doesn't mean that management and the board of directors should sit back and wait for the suggestion box to be filled and then act. It doesn't mean that they should or would act once it's filled. It means, however, they are alert to any and all possibilities.

Nothing of any substance at all would be done without some sort of feasibility and market testing of an idea, a product, or a different way of doing something. This suggests appropriate feasibility studies of varying degrees of sophistication.

This also suggests a research posture with adequate capacity in personnel and facilities to conduct research of whatever nature is needed. Perhaps having an attitudinal bent toward appreciating the importance of this aspect of measuring performance is sufficient. Perhaps a working relationship with university people or other outside consultants would be most feasible in many cases. The necessary ingredient, however, is a constant concern with whether there are better ways to do what we're doing and if some of the things that are being done now should be eliminated completely or perhaps replaced by others. With the requisite attitudinal attributes, progressivity on the part of the cooperative is very likely to be assured. Again, conduct options impacting upon progressivity most likely fit into the management box.

EQUITY

Equity, with its connotation of fairness, is perhaps the most slippery concept we are using. Its slipperiness, as expected, makes proxies for use in measuring this dimension of performance perhaps less readily available than for any of the other dimensions.

On the other hand, cooperatives with a group action and working together bent may be best able of any form of business organization to grope for handles on possible measurement vehicles.

Member refunds based on patronage and per unit capital retains based upon member involvement are fundamental cooperative principles, so there's no within-cooperative problem with these as principles or policy. However, the actual programs used to implement the policy may present problems at times.

That cooperatives should be financed to the fullest extent possible by those who are currently using it—the so-called currency rule—is generally accepted as an operating procedure. Under most circumstances, it could be construed as equitable for this to be the case and perhaps inequitable if it is not.

At times, however, cooperatives have not established equity redemption plans which assure fulfillment of the currency ideal. Redemption plans that reasonably fulfill this requirement in those cases where members have retired, died, or for whatever reason no longer are members of the cooperative should be devised. Other areas within the cooperative such as uniformity of contract provisions relating to acres or other units involved and adequacy of information—communications systems for all members—might be mentioned. These could be used as proxies for measurement of this dimension if tailored properly.

Another area for possible use relates to the guarantee of market provision in the contract of many cooperatives with their members. As proprietary firms choose to close facilities or move them to other areas, cooperative management is faced with decisions as to how markets can be provided for those members who had been shipping products to that firm(s). This may cause heavy strain on the ability of the cooperative to carry out this obligation and may even be temporarily inequitable to other members. It is an essential feature of cooperative equity, however, and over time adequate programs are in all members' interest. This may even have a positive performance impact outside the cooperative in that the public interest is being served by the cooperative stepping in to keep the members operative who had lost their market.

Various forms of evaluation procedures might be used to determine the degree to which the equity criterion as a performance dimension is being met. Number of members faced with loss of markets and what loss was averted would be such a measure. This again would be a conduct area formed as a result of board of directors' policy and implemented programs designed by management.

MEMBER SATISFACTION

Our last area established as a performance standard is member satisfaction with the services being received from the cooperative.

This involves much more than concern with the services or needs which brought about the formation of the cooperative in the first place. There should be a continuing concern with whether those needs have changed or whether the service(s) is no longer needed. In addition, there should be a continuous search for ways in which these or any other services might be provided more effectively and efficiently.

Communications with members is considered one of the most essential areas in the area of member relations and reaching the goal of having member satisfaction. Nothing can overcome a situation in which the economic needs of the members aren't being satisfied. A feeling on the part of the members that they do not really belong and are not being heard can bring dissatisfaction in the long run, even if economic requirements for services and various functions are being met satisfactorily.

An area that can pay greatest immediate and also long-term dividends in member satisfaction is effort expended in training and developing young leadership potential. Stimulating an interest in being involved in the cooperative on the part of young members can be very helpful in member satisfaction in the current period and in assuring that the cooperative will be in good hands when it is necessary for new leaders to take charge.

MEASURES TO USE

Various measures of member satisfaction can be used. An implicit measure is the degree to which the cooperative is being adequately financed. The quality of and the positive–negative nature of discussions at local, district, and annual meetings can also be a barometer. The types and content of resolutions presented at various places serve as indications of the thinking of the members and reflect their satisfaction or concerns. Periodic sampling of member thinking in regard to various functions being provided by the cooperative, their sources and uses of information, and their suggestions for action and functions of a differ-ent nature can be very helpful. Even if they provide no information not already known, they serve the psychological purpose of causing the members to have a more positive attitude about the cooperative. The fact that the opinions and thoughts of the members had been sought can be very helpful.

Conduct options most apt to lead to sound decisions regarding the performance standard of member satisfaction can be centered in the area of membership relations. This does not mean, as we've stressed before, that there is not a horizontal mixing and mingling of all the suggested areas in the conduct option part of the framework. Areas within the management box, the area of financing, and the area relating to equity and fairness are certainly involved. There is some danger, it is granted, in setting up such an analytical framework, because the impres-sion might be gained that the parts stand alone and are not interrelated. Nothing of that sort should be allowed to enter our thinking, for nothing is further from the truth. All are involved in the process of trying to have or bring about a cooperative that serves its purpose. As is true in so many cases, a smooth-running and properly functioning cooperative reflects far more than a simple summing of what is done in this box and in that box. Probably nowhere is the potential of synergism greater than in agricultural cooperatives.

EVALUATION—PERFORMANCE MEASURING AGAINST OBJECTIVES

The aim in this chapter has been toward stressing the importance of evaluation. It has been strongly suggested that evaluation lacks purpose and meaning unless it is placed in a result of action or lack of action and a function measurement framework against stated objectives or stan-dards which were agreed upon. Once the performance criteria are accepted and action options are specified, they can become logical com-ponents of an analytical framework by using appropriate evaluative procedures. There has been no intent to suggest that the action options or the performance criteria used are the most correct ones or that they are the only ones. The process and the attitudinal prerequisites for an evaluative stance are the important aspects we've stressed in this effort. It has been said that assessment of the overall effectiveness of a coopera-tive rests in interpretation of outcomes in terms of strategic economic and social exchange goals that were established. That is what evaluation is all about.

It is difficult to take the position that any attempt, even though incomplete, at evaluation is better than no evaluation. There is the possibility of damage being done if, due to the incompleteness or faulty nature of the methodology or incorrect interpretations of the results, wrong decisions are implemented. It is difficult to discourage any attempts, however, because evaluation is a process encouraging its own refinement once it is diligently and purposefully undertaken and repeated.

BENEFITS FROM SYSTEMATIC EVALUATION

Systematic evaluation along the lines indicated and making use of the suggested framework can lead to these results, among others:

1. Make adjustment to change through utilization of new information, new technologies, and new processes.
2. When evaluation is done and done well, it upgrades the performance of all cooperative leaders.
3. It generates an early warning system. The quicker the management and board learn of and react to problems developing in any area, the quicker they will be in position to take remedial action. Merely looking at annual audits may be too late for taking appropriate action. A system of monitoring changes and assessing their impact along with suggesting remedial action should be developed by management and the board.
4. Systematic evaluation can suggest ways in which the cooperative can be made more useful to the members. New information, new technology, new processes, new plans, new services, and new skills needed must be examined, and feasible adjustments which they suggest should be made.
5. Evaluation and planning are complementary and related. Planning requires the examination of goals, strategies, actions, use of resources, etc. Evaluation is determining whether the targeted goals have been reached. As shown in our analytical framework, having both goals and plans is essential in the process of evaluation.
6. Systematic evaluation procedures properly conducted can point to problem areas and can indicate whether outside or external evaluations may be needed. Such areas as feasibility studies, various types of internal appraisals, complete business audits, policy regarding insurance and various forms of liability, relationship with government, and agricultural policies may be most properly and effectively handled in the external evaluation process.

SYSTEMATIC PROCEDURE IMPORTANT

Whether evaluations are aimed at potential internal problems and are conducted by the cooperative itself or whether external evaluation by outsiders is suggested, it is important that a systematic procedure is used. In all cases, it should relate goals and objectives to the degree to which they have been achieved. Evaluations should involve the conduct options that have been used in the process of trying to achieve established goals. In this process, performance of the cooperative is being measured against the goals and, implicitly, the effectiveness of actions that were used. Rather definitive positions can then be taken by management and the board regarding the effectiveness of the conduct options as they were implemented. Soundness of the decision to use whatever conduct options were used is thus tested. The effectiveness with which they were used is measured. Bases for improved and perhaps more soundly based decisions should be the product or dividend realized from this effort.

REFERENCES

Arason, G. S. 1984. Our Canadian Programs. NICE, Bozeman, MT.

Babb, E. M., Boynton, R. D., Lang, M. D., and Schraeder, L. F. 1981. Performance of Cooperative and Non-Cooperative Firms. NICE, Ft. Collins, CO.

French, C. E. 1981. Cooperative Performance versus Proprietary Corporate Performance, Related to the Cooperative Principles or "Cooperative Survival." NICE, Ft. Collins, CO.

Humble, L. 1981. How's Your Cooperative's Performance—Compared to What? NICE, Ft. Collins, CO.

U.S. Department of Agriculture, FCS. 1982. Dairy Farmers' Evaluation of Northeastern Dairy Cooperatives. ACS Research Report No. *19*.

TO HELP IN LEARNING

1. Quiz a number of your peers regarding what areas, if any, in which they are concerned with evaluation. Relate their response to their major academic area. What do you conclude?

2. Consider some hypothetical activity. Evaluate its performance. Give details.

3. Conduct a 45-minute seminar–discussion with your peers on the topic: "Evaluation—Its Purpose and Methods Used."

4. Discuss with a cooperative member and determine the extent of the members' knowledge regarding evaluation procedures used by the cooperative.

5. Talk with the general manager of the same cooperative and determine management's attitude regarding evaluation and processes used and what procedures the cooperative uses.

DIRECT QUESTIONS

1. What are performance criteria?

2. If you were a cooperative manager, how would you know if your cooperative was performing satisfactorily?

3. Evaluate the framework suggested in this chapter for evaluation purposes.

4. Again, what is a proxy? How are proxies of special value for our purposes here?

5. Why did we start with considering performance first?

6. Define efficiency.

7. What does progressivity mean to you?

8. Define equity.

9. What does the phrase, "be in an evaluative mode" mean to you?

10. Does the use of boxes in our framework mean that each element fits neatly into its own area? Comment.

11. Define conduct. What are conduct options?

12. What is meant by a "good citizen" concept?

13. Input–output ratios measure what type of efficiency? Are there other kinds?

14. What is pricing efficiency?

15. What is social efficiency?

16. Where does social efficiency show up on the balance sheet and income statement?

17. What is the purpose of research?

18. How would you measure progressivity of a cooperative?

19. Fill in the blank. If evaluation lacks purpose, it _____.

20. React to this statement, "It is better to have tried to evaluate and failed than not to have tried at all."

TYING-TOGETHER QUESTIONS

1. Would you say it is easier to evaluate the performance of a cooperative corporation than a private corporation? Support your position.

2. Is what constitutes "success" the same for a cooperative corporation as for a private corporation? Explain your answer. Is success the same for all cooperatives? Explain.

3. Would you say that, generally speaking, it is easier for a cooperative corporation to achieve success than for a private corporation? Explain your position.

4. Would you say that a cooperative corporation is more likely to perform well in the area of progressivity than a private corporation? Explain your position.

5. Which type of corporation, cooperative or private, would be more concerned about equity and more likely to perform well in relation to this criterion? Explain your position.

18

Agricultural Cooperatives— Looking Ahead—Issues and Challenges

Agricultural cooperatives have been portrayed as a form of business enterprise in a market economy which is particularly adapted to serving the needs of the agricultural sector because of its structural arrangement. Cooperatives are designed to provide services to agricultural producers which they, as individuals, cannot provide for themselves or cannot do so as effectively.

Their uniqueness, tied very closely to the unique characteristics of agriculture itself, lies in the owner, management, user, and reason-for-belonging areas. These features make of cooperatives a distinct form of business organization. At the same time, they are confronted with much the same business practice requirements as are other forms of business enterprise in our society. Their simply being a cooperative is not sufficient—they must be soundly financed, managed, and run if they are to be viable.

"SPECIAL" CONSIDERATION

Because of its unique nature in the areas just mentioned, the cooperative corporation has been given what some may construe as special considerations. One of these is in the area of financing.

Since cooperatives are formed for the purpose of providing services to their members and membership is closely tied to this need, sources of financial support are needed which reflect their particular requirements. Their needs are grounded in the unique characteristics of agriculture in that particular types of credit are needed that have repayment schedules tailored to agriculture's uniqueness with respect to seasonal, short-term, and long-run considerations. Members are expected to finance a significant share of their cooperative's needs even though they are not members as investors, as is the case with investor-oriented corporations. They do need, however, to have funds from external sources available to them for a portion of their financial requirements. This requires special forms of lending agencies that recognize the uniqueness of their clients.

SPECIAL QUESTIONS

Perhaps because of these special needs and requirements, special questions have arisen. An example is within the area of taxation. If cooperatives were dependent in a special way upon their members for financing, should they be penalized by assessing a patronage tax on them if their members elect to leave their patronage allocations with their cooperative for capital purposes? The single tax concept in which a tax is collected either from the cooperatives or their patrons was devised to recognize this special cooperative arrangement. It was recognized that cooperatives needed to be capitalized by their own member–patrons and that use of the single tax concept was a feasible way for them to do it.

INDUSTRY STRUCTURE

Finally, industry structural arrangements, as related to numbers of firms within an industry and the sizes of those firms, were examined carefully in Part I of this book. In this inquiry, it was pointed out that in those industries such as agriculture, made up of a large number of small producers, the individual firm (farm) was powerless in that it had no impact whatsoever on prices or terms of trade in the market. This was tied to the fact that offerings of products by these individual small producers was such a small part of the total on the market that its impact was negligible on total supply and, of course, on market price.

On the other hand, industries such as those supplying inputs to farmers—fertilizer, feed, chemicals—are structured far differently than agriculture. Because these industries are made up of a small number of large firms, the individual firm is not helpless in the marketplace. Its actions are not independent of the actions of all other firms. Firms in this structural configuration can set prices, control supply, differentiate their products, affect demand, and so on. These are conduct options that are not open to individual farmers because they are part of an industry made up of thousands of firms (farms). The input suppliers are price makers and farmers as individuals are price takers.

The same situation holds on the buying side with respect to those who buy output from farm firms. They too can exercise conduct options that are available only to firms in monopolistic or monopsonistic positions.

An interesting and significant conclusion reached in discussing the implications of structure to economic power of individual firms was that firms within industries that are characterized by small numbers and largeness had the potential to establish goals and take actions which lead to outcomes in which their, the firms', interests are served, but the public interest may not be served. In other words, there could very well be many situations in which the public interest would not be served.

In the case of industries structured as is agriculture, however, which are more subject to the invisible wand phenomenon set forth in Smithian theory, the goals of the firms are much more apt to be in accord with those of society.

This line of reasoning was followed in the deliberations that led to passage of our antitrust legislation and to the Capper–Volstead Act which made it legally possible for groups of farmers to join together in jointly marketing their products without, per se, being in violation of the Sherman Antitrust Act of 1890 or the Clayton and Federal Trade Commission Acts of 1914. The unique characteristics of agriculture were recognized and legislation was designed that would take those characteristics into account. Goals of farmers through their cooperatives and goals of the general public were held to be in very close alignment, so it was in the interest of the general public that the cooperative corporation form of business enterprise, along with special arrangements in the areas of financing, taxing, and so on, be sanctioned.

Despite the fact that the Congress at various times has recognized the public interest posture of agriculture by passage of legislation reflecting this recognition, the concept of the agricultural cooperative form of business has not gone unchallenged. In many cases, it has been severely criticized. Let us now review some of the areas in which agricultural cooperative corporations have been subjected to criticism.

ISSUES AND CHALLENGES
RELATING TO COOPERATIVES

The purpose of this chapter is to set forth some of the issues and challenges in sufficient detail that the student of cooperatives might be encouraged to study the issue further with a view of becoming sufficiently knowledgeable that a position regarding it might be established and articulated. While many of the issues that will be mentioned will be rather straightforward and are readily apparent, such as that of the single tax, most will be couched in a public interest framework. By this is meant that a procedure is encouraged in which the argument is not one of cooperatives vs noncooperatives, but an examination of rules and regulations relating to cooperatives in which the objective is to determine if those regulations bring about economic activity in the public interest. This procedure is suggested because it has the potential of encouraging more constructively critical analyses. Definitive positions may not be reached in each and every case, but the process of trying to reach one will be more interesting and constructive than might be the case with other procedures. To the fullest extent possible, this more basic procedure of constantly asking the question, "Does this regulation, law, or rule implement policy which is in the public interest?" is suggested even when rather pedestrian issues are posed.

The issues and challenges to be covered will include those that are basically operational in that they are internal to the cooperative and those that are external to the cooperative. This breakdown can be made by the reader in the interest of simplification despite the fact that most of the issues and challenges have implications for both. An attempt will be made to handle each issue in a manner that creates interest in pursuing it further with the hope that a comfortable position will be reached. The aim will not be to push toward alignment with a certain position, although the discussion in some cases may suggest such an aim.

THE PUBLIC INTEREST

Since we've stressed the importance of some degree of conformity between the goals of a business enterprise and those of society, and since we have suggested that the issues and challenges be handled within a public interest context, we should try to make what we mean by public interest a bit clearer.

It has been said that there probably is no way to properly define public interest, since there are so many segments of the public and so many interest groups. It is true that there are many public segments and many groups have special interests which they try to pursue, many times at the expense of other groups. However, there must be interests that go beyond a narrow perspective which can be described as public interests when the term public in this usage is broadly interpreted to encompass all the other public segments. Overall public interest objectives in this sense could be fairly well captured by the performance objectives we established in Chapter 17 for business enterprise.

A public policy has been defined as a set of formal rules with the force of law, promulgated and enforced by the public through legal institutions as a means to achieve public, social, and economic goals. In the context in which we are studying cooperatives and in which we developed the bases for the set of formal rules relating to this arrangement of business enterprise, it seems appropriate to suggest that the question that was posed previously should be constantly kept in mind. Perhaps it should be restated—Does this regulation or law relating to cooperatives implement policy which results in performance that is in the public interest? Even if we fall short of adequately defining public interest, this focus upon what is seemingly good for all of society should be helpful in our considering the challenges and issues which are set forth. Let's now consider each one.

OPERATIONAL ISSUES— INTERNAL TO THE COOPERATIVE

Constant Education

One of the so-called Rochdale Principles is member education. While it perhaps lacks the basic, fundamental nature required of a principle or a law, it must be considered a practice which cooperatives can neglect only at great risk to themselves. This is because cooperatives are a special type of business organization requiring a special effort in the areas of communications and education, and this is close to being a necessity. In 1966, the International Cooperative Alliance stated this practice as follows:

> All cooperatives should provide for the education of their members, officers, employees, and the general public in the principles and techniques of cooperatives.

Despite the fact that cooperatives in general say that they believe education is very important, many make little, if any, effort in following through with their indicated belief.

In addition to reaping the benefits from a well-informed membership in areas such as loyalty and meaningful involvement, there is also much to be gained from extending educational efforts beyond the cooperative membership. Over the past decade or so, agricultural cooperatives have been besieged by an unprecedented series of criticisms. Tax issues and the antitrust exempt status of the cooperatives have been challenged and criticized.

It may well be that such criticism stems from a lack of understanding of why the Capper–Volstead Act, which provides for exemption from provisions of the Sherman Antitrust and other Acts under certain circumstances, was enacted. Critics may fail to grasp the significance of the position that those who enacted this legislation felt that they were providing for the implementation of legislation that would ultimately be in the public interest. Constant education is needed in this critical area if understanding on the part of the general public is to be achieved.

The lack of adequate education becomes more difficult but more important as cooperatives grow larger and lines of communication become longer. If member concerns are to be heard and addressed, if their attitudes about their role in the cooperative are to be positive, and if their understanding of pricing, capital programs, equity redemption plans, and the like are to be sufficient to sustain a healthy organization, the place of constant education must be established as a policy and implemented with innovative and effective programs.

Another basic group, perhaps it too could be called a public segment, with which cooperatives should show special concern in constant education efforts is the potential young cooperative leadership group.

European cooperatives have been said to take the position that a cooperative without an education program will last a generation and a half. Many cooperatives are now faced with that last half-generation problem. The task of educating young and newer members and developing leadership is of utmost importance. They must be apprised of why the cooperative was formed and what is necessary for it to continue performing its role satisfactorily.

Older generation cooperative members and leaders have a special obligation to pass on, in addition to wealth and assets of a tangible nature, their drive, their sense of purpose, and the memory of the effort they put forth in bringing the cooperative into being. Only in this way, through an effective education program, can they avoid the succeeding generation problem.

To the extent that cooperatives serve the public interest and since they must serve their members if they are to remain viable, it seems imperative that cooperatives should be constantly engaged in educational efforts. This is a major challenge.

Planning

We have moved away from the position held by some that the cooperative represents a way of life and that just being a cooperative is sufficient reason for business success. While there may be a basis for the position that in addition to providing services in an economically efficient and effective manner which are not feasible from an individual standpoint, there are fringe benefits that come from working together toward common goals, there is nothing to suggest that just being a cooperative is sufficient unto itself. There must have been a need for the cooperative in the beginning, it must have been formed in accordance with appropriate procedures, and it must be soundly managed by the manager, board of dircetors, and the members. This is the only route to a bottom-line position making it possible for the cooperative to continue.

One of the actions on the part of management and boards of directors which suggests an understanding and appreciation of this necessity, a satisfactory bottom-line position, is in the area of planning. This action in both short- and longer-term planning is a healthy indication of the movement away from a complaint–treatment posture on the part of cooperatives to one of health maintenance. This is an anticipatory concern with potential problems before they arise or certainly before they become full blown. It also reflects the position which has become prevalent that cooperative corporations are a unique business enterprise form in many ways, but they are subject to the same pitfalls stemming from poor management and the failure to look ahead as any other form of business. Let's confine our efforts in this challenge area to considering the rationale for and what is involved in long-run planning.

What Is Long-Run Planning?

Long-run planning is a comprehensive, coordinated, purposeful development of relevant data regarding the cooperative and the economic environment within which it has operated in the past, is now operating, and will probably be operating in the future. Once relevant data are developed, they are organized in such a way that they may be interpreted meaningfully in relation to the goals the cooperative has established for itself. Here again, we are reminded of the importance of an analytical framework which provides guidelines for such efforts.

What are our economic and other relevant environments and what are our goals and objectives? Given these, what are the conduct options that are open for use and that are most likely to result in the objectives being achieved if pursued?

A question arises as to how long long-run planning should be. It is longer than day-to-day or what is usually recognized as short-run planning. There is no definite answer, however, as to how long is long. The appropriate length of time is the period during which major policy decisions or technological changes may have their greatest impact.

An example would be in a situation in which a cooperative is considering its future role in balancing supply and demand of its product in a market. Also involved is the cooperative's guarantee of market for the raw product of its members. What, if any, additional facilities would be needed in order to carry out its objectives under the most likely scenario?

Obviously, this is long-run planning of a most strategic nature. Data and information needed relate to long-term supplies of the raw product being considered, demand for the processed products which may be appropriate, and the costs involved in developing the facilities needed on the basis of the data and information generated. Directly involved as a definite part of all this is a sales and marketing plan (probably the most important part, and this should be done first using reasonable assumptions), a facilities plan with location and the like, an organization plan designed to facilitate and coordinate all efforts, and a financial plan developed for an appropriate period. Implicitly involved is the need to keep the members appropriately and adequately informed, since there is no way any plan will work without the understanding and contributions that have to come from the members. Keeping employees properly and adequately informed is also a part of the overall strategy.

Once the data and information have been mobilized and placed into a form in which they can be interpreted, the board of directors and management are in position for making a decision. All the bases have been touched and a decision to go, delay, or not go is in order.

If the decision is to go, the next step is implementation of the plan. This is not an easy task, but if all bases have been touched and all—the board, management, members, and employees—are committed to it, its chances of having been soundly based and properly implemented will have been enhanced.

Long-range or any type of planning will not eliminate risk and uncertainty from a business, nor will it assure success in achieving objectives. It is, however, the best way of selecting the scenario(s) that has the greatest likelihood of success and thus the least risk and uncertainty.

Actual results will no doubt deviate from those expected. Constant observation is needed and perhaps in-course corrections in plans suggested on the basis of data and information updating, which should be done on a systematic basis, may be in order. The sooner the deviations from what is expected are noted, the sooner corrective actions can be taken. Deviations can be prevented in many cases and those that occur can be held to manageable size. Properly conducted purposeful planning is a necessity.

Evaluation

A basic and very complementary part of the planning and implementation process just discussed relates to the process of evaluation. With goals or objectives constantly in mind, with planned updating of information and data bases, and with maintaining a "how are we doing" posture, the evaluation process becomes an integral part of the planning processes. This is true with the annual report process, for example, in short-term planning and is true in our concern with deviations from expected results in the longer-term planning process.

Legal Foundations

Certain legal statutes must be complied with by cooperatives if they are to be provided an opportunity to operate in a business-like manner in meeting the needs of its members.

A group of producers who wish to form a cooperative corporation must comply with the applicable state statute in which they wish to operate. Basic documents including the Articles of Incorporation and bylaws must be drafted to meet statutory requirements of the relevant state. Marketing contracts or agreements may be needed, especially marketing cooperatives using pooling. These must be drawn in conformity with the bylaws of the cooperative.

Other state laws which may be relevant and about which it is the responsibility of management and the board of directors to be aware and to observe may include (1) basic contract law, (2) the uniform commercial code, (3) banking and insurance laws, (4) workmen's compensation, (5) unemployment insurance, (6) securities laws, (7) state tax laws, and (8) all other laws and regulations under which a business operates.

Legislation at the federal level, in addition to that which has been discussed, about which cooperatives should be informed include (1) the Packers and Stockyards Act of 1921 relating to refunding any portion of charges by stockyard operators; (2) the Grain Futures Act of 1922, stating that cooperatives could not be excluded from membership and that cooperatives could refund any net margins realized in conducting the business of the cooperative; and (3) the Cooperative Marketing Act of 1926 providing for a division of cooperative marketing within the U.S. Department of Agriculture to render services to cooperatives. The Act also authorizes agricultural cooperatives to acquire, exchange, interpret, and disseminate past, present, and prospective crop, market, statistical, economic, and other similar information by direct exchange between such persons as producers of agricultural products and/or associations or federations of producers or through a common agent created or selected by them.

A federal statute which may attract the attention of those interested in public policy and its implementation relates to our securities laws.

The sale and exchange of a security interest in an interstate business is under the jurisdiction of the Federal Securities and Exchange Commission. The Act requires registration of a security issue before it can be sold to the public and its purpose is to assure full disclosure of information needed by the buyer of the security to make an informal decision in regard to purchasing it. Failure to register involves heavy penalties.

Registration of securities under this Act generally applies to all security issues, but is not required of cooperatives that qualify for tax treatment provided under Section 521 of the Internal Revenue Code. Antitrust provisions apply to all corporations, cooperative or noncooperative, but questions have risen from time to time in this area that may have significance for agricultural cooperatives.

There are perhaps other legislative and regulatory provisions at the state and federal level that have relevance for agricultural cooperatives. The challenge to cooperative leaders is to be informed about them, comply with them, and take them into account. There is the further suggestion that the mere fact of being an agricultural cooperative is not sufficient. Special competence in being familiar with all relevant regulations such as these, the preparation of and changes in the cooperative's bylaws, and other areas with legal overtones is required in most cases.

To Merge, Consolidate, Federate, or Integrate?

Cooperative corporations, just as other types of corporations, are faced with decisions regarding expanding operations and how it can be accomplished most effectively. Two methods are available for use once the decision is made to grow—grow internally by expanding facilities, services, and hopefully the ability to render services more effectively and more profitably, or grow by combining with its operations the activities of a separate business. The same basic business decisions are involved whether a cooperative is considering growth or increased size by merging, consolidating, federating, or integrating, either horizontally or vertically. The soundness of the decision regarding the need to get larger is critical, and once that decision is made, the method, whether internally or by following one of the routes previously indicated, then becomes crucial.

It is quite apparent that we must return again to the area of strategic planning—this time of a long-run nature. To get big just to get big is hollow. People today are concerned about bigness, yet there is perhaps an underlying feeling that cooperatives have as much incentive and right to grow as do other businesses.

The first step is the development of a long-term strategic planning model of the type suggested previously. Basic goals and purposes must be identified in terms of today's and tomorrow's needs and then organized to meet these needs. Future members, their needs and their means of meeting them, must be identified. A basic challenge is to adhere to our cooperative principles as we proceed to get bigger by whatever route. Mergers usually suggest a greater degree of centralization, so how do we involve our local farm leaders in the decision framework if we move to a larger organizational arrangement?

Business efficiencies may well be achieved by getting larger through scale economies. Per unit costs may be reduced. Overlapping membership may be eliminated, more effective management may be afforded, and economies in transportation and other functions may be realized. There may be instances where relatively simple coordination of functions such as pricing over larger areas, with uniformity of prices being charged for packages of services being provided, is needed. This may suggest a federation arrangement rather than merger or consolidation into larger cooperatives.

In considering vertical integration by any means, it is well for cooperative leaders to keep in mind that the further the cooperative is from the ultimate consumer, the weaker is its market power. Access to markets for members' products may be achieved by contractual arrangements by the cooperative with processors and by overseeing the terms of the contracts in such a way that producer–members are not disadvantaged. Complete vertical integration forward to the consumer would also provide such access to markets. It is well to remember that those who control exposure of a product to the ultimate consumer have the greatest market powers.

The challenge to cooperative leaders is to determine to the best of their ability what their future member needs are and gear up in terms of size and organizational arrangement to meet those needs. Only such soundly based reasons can justify getting bigger by any route. The route to follow must also be equally soundly based. The issue is to do whatever sound planning dictates in accordance with recognized basic cooperative principles and procedures.

Joint Ventures

The key legal question arising from joint ventures in which business arrangements are worked out between cooperatives and other organizations is, "Who is a producer?" It is remembered that in the key language of Section 1 of the Capper–Volstead Act, "Persons engaged in the production of agricultural products as farmers, planters, ranchmen, dairymen, and nut or fruit growers may act together in association, corporate or otherwise, with or without capital stock in collectively processing, preparing for market, handling, and marketing in interstate commerce such products of persons so engaged." This language quite clearly provides cooperatives an exemption from federal antitrust laws. Court decisions have held that the exemption was not absolute. For example, the Supreme Court ruled in 1939 in the Borden case that when a Capper–Volstead cooperative enters into an agreement with a business that has no antitrust exemption, the cooperative is to be treated the same as any other corporation. The Court also ruled in 1967 that Sunkist Growers, Inc. had no Capper–Volstead immunity because some of its members were packing houses and not producers.

In general, it appears that Capper–Volstead confers very little immunity from the antitrust laws to a cooperative when it enters into agreements with noncooperatives for various functions. They are not producers. There are many unanswered questions in this area. A major challenge to cooperative leaders is that of understanding how to use the Capper–Volstead exemption to the fullest extent in serving its members, but to remain completely within the confines of the exemption provided by the law.

Undue Price Enhancement

As recalled, Section 2 of the Capper–Volstead Act empowers the Secretary of Agriculture to take action against a cooperative which monopolizes or restrains trade to such an extent that the price is unduly enhanced. The language of the Act specifies that if the Secretary shall have reason to believe that the price of any agricultural product has been unduly enhanced, the association shall be served a complaint stating the charge and requiring that the association show, within a specified time, why an order should not be issued directing it to cease and desist from the alleged practices that were bringing about the prices alleged to be too high or unduly enhanced.

The intent of the language is quite clear—there is no absolute exemption from federal antitrust legislation in the area of pricing, and if prices are deemed to be unduly high, then the Secretary shall take steps to see that whatever the cooperative might be doing to bring about such prices is stopped. But the problem is that no one has ever defined undue price enhancement. This, then, constitutes a major policy issue and a challenge for cooperative leaders.

Several cases have arisen where the Secretary took steps to use the authority provided under Section 2 of the Capper–Volstead Act. An important and fairly recent case of this nature evolved when the Secretary established a blue ribbon committee to determine if prices were unduly enhanced by dairy cooperatives who were charging superpool prices, prices that were above the federal order minimum prices.

The position was taken by those who asked that an investigation be conducted that if negotiated superpool milk prices were 50¢ per hundredweight or more above the federal order minimums, then this constituted undue price enhancement. Information and data were supplied by the cooperatives involved and a determination was made in regard to the extent to which the negotiated prices were above the producer minimum prices announced under the various federal milk marketing orders. The necessary information was provided to the Secretary of Agriculture. The Secretary ruled that even though prices were as much as 50¢ per hundredweight above the announced minimum prices, this did not constitute undue enhancement because the premium charged was actually a payment to the cooperative for performing marketwide supply–demand balancing services for the entire market. It was ruled that this was a necessary service in order for the market to work and that there were costs involved in rendering it. It was further ruled that the number of dairy farmers had continued to decline, and this suggested that prices were not too high.

As indicated previously, this constitutes a major policy area. No one has ever defined what constitutes undue price enhancement. One group, the National Commission for the Review of Antitrust Laws and Procedures, asked Congress to come up with a definition.

The Capper–Volstead Act does provide some guidelines. It states that for a violation of Section 2 to occur, acts of monopolization or restraint of trade must be committed by the cooperative and must be the cause of the undue price enhancement, but restraint of trade is not defined and monopolization is not defined. These issues have been dealt with on various occasions, so some of the boundaries are known, but much remains unclear.

Most interpretations of the intent of the Congress when it passed the Act in 1922 and its legislative history are that Congress intended that agricultural cooperative marketing associations be permitted to legally raise prices received by the cooperative members above the level that would have existed if individual farmers were pricing their product, but that there is, in fact, a level above which prices are considered to be unduly enhanced. That precise level has not been defined.

This lack of definition of what constitutes undue price enhancement may leave some a bit uncomfortable, but it appears to be in perfect harmony with our usual procedures under the democratic process. The decision that the acts of a cooperative constitute monopolization and/or restraint of trade and result in undue price enhancement must be reached on a case by case basis. This involves the use of the rule of reason and represents a distinct departure from the, per se, illegal position taken by the courts in early interpretation of the Sherman Antitrust Act.

Reaching a decision regarding whether prices have been unduly enhanced or were unreasonably high in a particular case would involve several steps. A definition of the economic market in which the cooperative operates would be necessary. The structure and conduct of the firms on both the buying and selling sides of the market would be analyzed in an attempt to gauge the market power of the buyers of agricultural products in the market and of those who provided inputs to producers in the market. Concentration ratios and market shares would give some indication of market power, but further analysis may be needed. In general, it might be expected that if it is found that the cooperative has no more market power than those who purchased its output or sold inputs to it, no judgment that undue price enhancement exists in the market could be rendered. Again, however, it is difficult to generalize even from this.

Any concerted, well-conducted study of the law and economics as they relate to Section 2 of the Act should make a positive contribution to our understanding it. It should be helpful in development of a more definitive policy in regard to its enforcement. As in all matters, however, the final decision in this regard will be made in our courts. When we recall that circumstances are never exactly the same in any two situations and that there are rarely, if ever, the answers to important questions, perhaps we may find what appears to be vagueness by design to be more acceptable. In any event, interpretation and enforcement processes involved in this section of the Act constitute a major policy issue challenge on the part of cooperative leaders.

Finance, Accounting, and Equity Redemption

Since our focus in this chapter has been on those areas which are most likely to receive attention from a public interest and public policy standpoint and which could result in policy and program rules for implementing the policy, it seems that financing, accounting, and equity redemption plans should be considered.

Various methods used by cooperatives to obtain capital may be subjects of public policy and implementation rules from time to time. Such methods as per unit retains or retention of patronage refunds are in this category. Cooperatives do sell stock but, as we saw earlier in this chapter, they are not subject to the Federal Security statute relating to registration and disclosure if they meet Section 521 requirements under the Internal Revenue Service Code. This issue may become sensitive to public concern.

Policy issues relating to equity redemption or nonredemption have arisen from time to time. Continual monitoring on the part of cooperatives is suggested. All areas of cooperative financing instruments are subject to some degree of sensitivity to public concern. In addition to finance, per se, being involved in these issues, there are also involved member rights and responsibilities with respect to cooperative financing, rights of financing members when they leave the cooperative for any reason, member bankruptcy, and cooperative bankruptcy.

In the case of accounting, there may be member rights involved in accounting procedures that do not take into account changes in the general price level, and inflation, stability, or deflation in depreciation schedules and, because of this, these procedures may reflect an incorrect picture with respect to member equity, patronage refunds, and the like. This area may not be as sensitive to public concern as are others, but it is an area of relevance.

Research

Cooperatives, as much as any type of business enterprise and perhaps more, should be research oriented. This is because many researchable areas have particular interest to cooperatives, since they fall into the area of public sensitivity which may result in policy formulation and implementation measures.

As in any case, cooperative policy-related issues may be placed into time categories and identified as current issues and longer-term or policy-anticipation issues. The first category, of course, would include research designed to satisfy immediate information needs relating to operation of the cooperative or to overall cooperative policy. The longer-term approach would be taken to identify, by anticipation or prediction, issues that may arise in some planning horizon specified by the cooperative.

The advantage of researching current or short-term issues is that of satisfaction in meeting immediate needs, at least partially, and it is usually addressed to a single issue. It is often done, however, under severe time constraints and in some cases, many parcels of relevant information may not be provided. Little analysis is possible. The most serious flaw, however, may be that problems are usually being addressed which are of immediate importance, suggesting that the cooperative leaders have adopted a reactionary rather than an anticipatory stance.

Longer-term research also has advantages and disadvantages. Given knowledge gained from past experience with policy issues along with factors with which they were associated, it is possible to speak in an anticipatory fashion regarding what research suggests as problems that may emerge in the future. With time leads involved in longer-term-oriented research, better research design can be used and modifications can be made. There exists the ability to act on the basis of relevant findings with the satisfaction that the decision makers were not caught unprepared.

Disadvantages, for the most part, relate to the accuracy of the selection of the events to be researched and their suggested impact or importance.

Another approach that may be used is one in which particular issues, either short- or long-term, do not provide the focus. It is aimed at generating data and kinds of information most likely required for a number of issues that might arise. While a great deal of information will be compiled which may ultimately be used, this approach lacks the interest and motivation provided to the researchers by the knowledge that real and/or potential issues are being researched and that efforts are being expended to shed light upon them.

The basic focus in this challenge area, however, is not to play up the importance of short-term, longer-term anticipation, or information-based research. Each has its place, and a judicious mix is suggested. The challenge to cooperative leaders is to have reliable information readily and potentially available upon which decisions can be based. This requires a research orientation.

Cut Ties with Basic Cooperative Principles

There are those who seem to take the position that cooperatives should change significantly or even cut all ties with basic cooperative principles and become regular proprietary business organizations. There are also those who would not suggest a complete break with cooperative principles and practices, but do suggest a break with some of them. In either case, the suggestion has rather significant implications not only for cooperatives themselves, but for other forms of business enterprise and perhaps for the public interest as well.

The first position suggested, that of breaking all ties with basic cooperative principles, has to be considered as being of a drastic nature. It would mean, of course, that there would no longer be an institutional arrangement of a cooperative nature. It would also mean that all the effort expended in attempting to understand agriculture, to determine what its unique features are, if any, and what they suggest as to how these features might be handled in the interest of agriculture and of the public, would be thrown overboard. The cooperative corporate business arrangement would be scrapped along with all the supporting structure, such as credit agencies. Implicit in such an action would be the position that structure of industries is meaningless, that agriculture is just the same as any other industry, and that there's no special public interest involved which justifies any special arrangement.

The possible impact upon the corporations that would be left is interesting to contemplate. There is little, if any, reason to believe or assume that in the process of eliminating cooperative corporations, the exact number that was eliminated would simply change hats and become proprietary corporations. Would this step automatically change the nature of agriculture and its peculiar needs and would these needs be served adequately by regular corporate structures? If agriculture didn't automatically change its stripes overnight, could its needs be served by traditional corporations operating in their traditional modes? These are interesting questions to contemplate, but it is easy to see that they are not easily answered.

Finally, what about the public interest? Would it be as well served as is the case at present in which a unique institutional arrangement, along with its supporting cast, has been designed over time in recognition of the unique features of agriculture? Cooperatives themselves are not the critical factor in these considerations. Overriding any kind of business organizational arrangement is the question of the public interest, its proper definition, and under what conditions it can best be served.

The second position is that of casting aside some of the basic cooperative principles and practices and moving partially in the direction of
being proprietary corporations. The areas most often mentioned relate
to those of one person–one vote, membership control, and requirement
that members of the board of directors be elected from farmer members.

Differences in the size of operations of members and the resultant
difference in the amount of business done with the cooperative provide
the basis for the voting issue. It is noted, however, that this area is
addressed in the Capper–Volstead Act in that the one person–one vote
or the limited returns on capital provision must be used by the cooperative.

The suggestion that farmers may not have the competence or ability
to perform their decision-making functions on the board of directors is
another matter. Their competence in their farm operations is granted,
but some argue that today's cooperative, so far removed from the Rochdale pioneers, requires knowledge and competence in so many areas
that it is not reasonable to expect that these can be found in farmers.

Those who argue against this position readily concede the point that
great knowledge is needed by the members of the board of directors.
They point out, however, that the board can avail itself of expertise
through the use of consultants, advisers, etc. This is the case with legal
services, economic advisers, accountants, and the like. They argue that
the role of directors in the member–board of directors–manager management trio is so vital to the success of the cooperative that it should not
be compromised by having directors who are not cooperative member–
patrons.

Judicious study of the issues involved in these areas and the establishment of positions on them based upon reliable information and data
represent a challenge to cooperative leaders.

Federal Income Taxation

Corporations operating on a cooperative basis are taxed under the
Internal Revenue Code in a manner that reflects their unique characteristics with respect to their reason for being, their ownership, and their
control. These characteristics and the taxing provisions stemming from
them relate to the return of net margins to users of the cooperative on
the basis of the extent to which they use or patronize it. Farmer cooperatives that are able to meet the rigorous requirements of Section 521 are
accorded two limited additional deductions from taxable income. As
covered in our previous discussion of taxation, these requirements relate
to dividend rates not exceeding a certain level, and business done with

nonmembers may not exceed business done with members. If these additional requirements are met, the cooperative is said to qualify for exempt status and may exclude dividends on capital stock from taxable income and may include in patronage refunds earnings from business which are incidental to their usual marketing activities, such as interest income. Despite interpretations by the Internal Revenue Service of Section 521, which have led to a steady reduction in exempt cooperatives, these are relevant areas relating to tax issues.

Everyone familiar with what is involved recognizes that this can be referred to as special tax treatment. Those who have sought its rationale in industry structural arrangements as related to market power and other unique characteristics of agriculture accept the special tax treatment status as being soundly based and justified. They further take the position that such treatment is in the best interest of the public because of the characteristics on the supply side in agriculture and because of the nature of the demand for the food products produced.

There are those, however, who take the position that this is not only special tax treatment, it is also preferential treatment. It is not accorded proprietary corporations and should not be accorded cooperative corporations, they argue.

These tax arrangements, as applied to agricultural cooperatives, are specific expressions of public policy concerning a specific type of business enterprise. There is no fundamental requirement that the Internal Revenue Service Code contain special provisions for taxing of cooperatives—it is an expression of public policy that the kind of tax treatment accorded exists.

The differences in the positions taken by the two groups, perhaps stemming from a very basic difference in the manner in which they view the issues, result in different outcomes in the argument and different positions being established.

One group concerns itself with the principles on which tax policy toward cooperatives is based. This implies, of course, a concern with the principles on which cooperatives are based and suggests a thorough examination and analysis of the structural arrangements of agriculture and those industries with which it has dealings on both the input providing and output buying side. The market power of all the actors, as individual firms, its extent, and sources, should be taken into account. Conditions on the supply side of agriculture relating to risk and uncertainty, its biological nature, and seasonal and cyclical characteristics involved, should be studied, and their importance in relation to the availability of an adequate supply of pure and wholesome food should be assessed. Study and analysis would then proceed to the demand side with the characteristics of the products being produced, their essential nature with respect to human needs, their demand characteristics as related to substitutes and the like would all be analyzed. All would be done in the context of the public interest. The outcome would be a position based on underlying fundamental considerations which would suggest principles on which tax policy toward cooperatives should be based.

Another approach likely to lead to a different position in regard to this issue is that of considering tax rules only as a description of how usual tax principles apply in the particular situation represented by a cooperative enterprise. This approach may very well focus on the tax provisions as they are applied to cooperative corporations and to other types of corporations and could very well result in taking a position that the provisions are applied differently in the cases of the two types of corporations, so there must be unfairness involved.

The challenge to cooperative leaders in this area appears to be fairly straightforward. Their arguments must be grounded in the principle area and buttressed by the type of analysis already suggested. The sometimes conflicting versions of cooperative principles should be resolved by the cooperative leaders, and care should be exercised in making sure that agricultural cooperatives remain completely within the bona fide Capper–Volstead category. Only if such a position is grounded in relevant principles and articulated in a meaningful way to all the relevant segments of the public can attacks be adequately addressed.

Once such positions are established and articulated, federal income taxation provisions in the Internal Revenue Service Code would be expressions of public policy as grounded in the public interest. It would not be left in the tax enforcement context of the Internal Revenue Service in terms of what it thinks a cooperative is. It is understandable that taxing authorities would attempt to resolve certain questions which may be raised in the area of tax collection. Tax policy toward unusual cooperative structures and arrangements may become more important as such arrangements become more common and more complex. The burden falls squarely on the shoulders of cooperative leaders to make sure that positions are soundly based on principles and that those principles are articulated in a professional manner to all relevant segments of the public.

Examples of where Internal Revenue Service concerns have found their way into tax enforcement practices include limitation of netting among parts of a cooperative, handling losses by a cooperative, farmer–member rules applied to Section 521 cooperatives, and issues of tracing income of a cooperative to patrons whose specific patronage generated the margin that was distributed. These are features of cooperatives thought by many to be beyond the legitimate concern of the Service for code enforcement purposes. Again, the burden rests on cooperative leaders to have their position grounded in sound principles, and all practices, acts, and the like should reflect those principles.

Mention has been made previously of the relatively high transaction costs borne by cooperative corporations because of the unique and strategic role of members in the management area and the resultant obligation of the board of directors and management to encourage their involvement in the affairs of the cooperative. Couple this with the semipublic nature of information and data relating to cooperatives and their meetings, and a mix results which places the cooperative in a much different situation with respect to efficiency of decision making, privacy, and privileged data and information than the private corporation. No answers are suggested, but this is an area of interest when issues are being considered.

Several areas have been covered under the heading of issues and challenges for the future of cooperatives. While other areas of significance exist, what happens in the ones discussed is of major importance to all cooperative corporations, noncooperative corporations, and other forms of business enterprise. This is also true in the case of all who have a public interest stake in our food system. That, of course, is everyone.

REFERENCES

Abrahamsen, M. A. 1976. Cooperative Business Enterprise. McGraw-Hill Book Co., New York.

Armstrong, J. H. 1981. An Overview of Past and Current Research Projects on Cooperatives. NICE, Ft. Collins, CO.

Bennett, J. W. 1979. Agricultural Cooperatives in the Development Process: Perspectives from Social Science. Monograph No. *4*. California Agricultural Policy Seminar, Department of Applied Behavioral Sciences, University of California, Davis.

Capper–Volstead Study Committee. 1980. Undue Price Enhancement by Agricultural Cooperatives, Criteria, Monitoring, Enforcement. U.S. Department of Agriculture.

Comptroller General Report to the Congress. 1979. Family Farmers Need Cooperatives—But Some Issues Need to Be Resolved. CED-79-106.

Comptroller General of the United States. 1979. Family Farmers Need Cooperatives—But Some Issues Need to Be Resolved. Report to the Congress, July 26, 1979.

Cooperative Future Direction Project. A Vision for the Future of Canadian Cooperatives. Cooperative College of Canada, Saskatchewan.

Flinchbaugh, B. L. 1981. It's Easy to Be Ignored If You Don't Have Your Act Together. NICE, Ft. Collins, CO.

Fogerty, A. J. 1980. Governmental Climate for Cooperatives in the 80's. NICE, Pennsylvania State University, University Park.

French, C. E., Moore, J. C., Kraenzle, C. A., and Harling, K. F. 1979. Survival Strategies for Agricultural Cooperatives. Iowa State University Press, Ames.

Harris, F. L. 1984. Building and Sustaining Member Commitment. NICE, Bozeman, MT.

Hofstad, R. 1976. Serving the Needs of Tomorrow's Cooperative Members. NICE, Virginia Polytechnic Institute and State University, Blacksburg.

Ingraham, C. H. 1979. Cooperative Issues, What's Your Position? Employee and Collegiate Seminar, Columbia, MO.

Jesse, E. V., and Johnson, A. C., Jr. A Proposed Monitoring System for Detecting Violations of Capper–Volstead, Section 2. ESS, U.S. Department of Agriculture; and University of Wisconsin, Madison.

Jesse, E. V., and Johnson, A. C., Jr. 1980. Marketing Cooperatives and Undue Price Enhancement, A Theoretical Perspective. WP-46, NC Project 117.

Jesse, E. V., Johnson, A. C., Jr., and Marion, B. W. 1981. Interpreting and Enforcing Section 2 of the Capper–Volstead Act. WP-51, NC Project 117.

Johnson, A. C., Jr., and Jesse, E. V. 1980. What Is Undue Price Enhancement? Paper presented at Western Agricultural Economics Association Annual Meeting, July 20, 1980, University of Wisconsin, Madison.

Kanel, D. 1982. Some Observations Based on Issues Raised in the Nine Workshops on Cooperatives, Small Farmers, and Development. The Land Tenure Center, University of Wisconsin, Madison.

Knutson, R., and Garoyan, L. 1980. Current Cooperative Policy Issues and Environment. NICE, Pennsylvania State University, University Park.

Leaven, C. D. 1984. We Can Do It Together, American Agri-Women. NICE, Bozeman, MT.

Manchester, A. C. 1982. The Status of Agricultural Marketing Cooperatives under Antitrust Law. No. 673. U.S. Department of Agriculture, Economic Research Service.

Mueller, W. F. 1974. The Economics and Law of Full Supply Contracts as Used by Agricultural Cooperatives. In Proceedings of a National Symposium on Cooperatives and the Law. University of Wisconsin—Extension, Madison.

Mullinix, P. E. 1979. What's Your Future in Cooperatives? Employee and Collegiate Seminar, Columbia, MO.

Nichols, J. R. 1980. University Relationships with Cooperatives. NICE, Pennsylvania State University, University Park.

North Central Regional Research Committee 117. 1978. Agricultural Cooperatives and the Public Interest. NC Project 117, Monograph 4.

Ratchford, C. B. 1984. A Prescription for the Future Role of Agricultural Cooperatives. Paper No. 1984-23. Department of Agricultural Economics, University of Missouri, Columbia.

Rieck, R. E. 1980. Cooperative Relationships with Post-Secondary Institutions. NICE, Pennsylvania State University, University Park.

Schomisch, T. P. 1980. University Emphasis on Cooperatives, Review and Recommendations. U.S. Department of Agriculture, Economics, Statistics, and Cooperative Service, Staff Report, Washington, DC.

Torgerson, R. E. 1983. Research on Cooperatives in the 1980s. Graduate Institute of Cooperative Leadership, University of Missouri, Columbia.

Turner, M. S. 1984. Searching the Proper Roles for Cooperatives and Universities, What Should Be the Cooperative's Role? NICE, Bozeman, MT.

U.S. Department of Agriculture, ACS. 1984. Advising People about Cooperatives. Cooperative Information Report No. 29.

U.S. Department of Agriculture, ES and CS. 1978. Organizations Serving Cooperatives. Cooperative Information Report 1, Section 5.

U.S. Department of Agriculture, FCS. 1974. Economic Development through Cooperatives. FCS Program Aid No. 1088.

U.S. Department of Agriculture, FCS. 1975. Joint Ventures Involving Cooperatives in Food Marketing. FCS Marketing Research Report No. 1040.

U.S. Department of Agriculture, FCS. 1979. Agricultural Cooperatives: Challenges and Strategies. Cooperative Research Report 9.

U.S. Department of Agriculture, FCS. 1980. Attitudes toward Cooperatives, by Bank Trust and Professional Farm Managers. ACS Service Report No. 1.

U.S. Department of Agriculture, FCS. 1981. Farmer Cooperative Publications. FCS Information Report No. 4.

U.S. Department of Agriculture, FCS. 1981. Is There a Co-op in Your Future? Cooperative Information *10*.

U.S. Department of Agriculture, FCS. 1981. Organizing and Conducting Cooperatives' Annual Meetings. ACS Cooperative Information Report No. *21*.

U.S. Department of Agriculture, FCS. 1982. Strengthening State Cooperative Councils. ACS Research Report No. *20*.

Weike, T. 1984. The Promotion of Democracy, Political Corollaries to Cooperative Development. NICE, Bozeman, MT.

TO HELP IN LEARNING

1. Contact the general manager of a cooperative and determine what is considered as being major challenges to cooperatives today. Get the reasons behind them, whether they are short-run or long-run, and the best idea of the manager as to how they will be resolved.

2. Contact the manager of a private corporation and determine what are considered to be major challenges to cooperatives today.

3. Examine the two positions expressed. Are they based upon principles or are they based upon differences between the two corporations? Explain fully.

4. Quiz a number of your peers in regard to how they view cooperatives in areas in which they consider them subject to challenge. Determine why they take their positions.

5. Prepare a short paper summarizing the major points covered in the above findings. Are the positions well founded? What do you recommend?

DIRECT QUESTIONS

1. Why are cooperatives being challenged or why should they be?

2. What is meant by public interest?

3. How could the goals of a cooperative and the goals of society coincide? Give examples.

4. Are private corporations being challenged too? Why or why not?

5. Is constant education a principle or practice? Why or why not?

6. What is long-term planning?

7. What is the basis of the saying that a cooperative without an education program will last a generation and a half?

8. Why is a satisfactory bottom line just as necessary for cooperatives as for other types of businesses?

9. How long is long-term planning?

10. Will long-run planning eliminate risk and uncertainty?

11. What is meant by the statement, "Bigger is better"?

12. What is vertical integration?

13. What is a joint venture?

14. Who is a producer and what is the relevance of this question?

15. What is undue price enhancement and from where does this term come?

16. Why are equity redemption plans so important to cooperative members?

17. What would be an example of a long-run issue? A short-run issue?

18. What is information-based research?

19. Which is most important, short-term, long-term, or information-based research?

20. What is an "exempt" cooperative?

21. How are evaluation and planning a part of the same process?

22. In your judgment, what is the single most important issue facing agricultural cooperatives today?

TYING-TOGETHER QUESTIONS

1. How would you defend a form of business organization on the basis of principle? State your principles and prepare your defense of some form.

2. How would you defend a form of business organization on the basis of comparison with other forms of business organization? Prepare such a defense in detail.

3. How would you determine in a professional manner whether prices had been unduly enhanced by a cooperative?

4. How would you determine whether the goals of a corporation and those of society are in harmony? Give details.

5. Prepare a short paper in which you develop fully the basic, underlying bases used by those who argue that sanctioning the cooperative form of business is, indeed, in the public interest.

6. Develop fully a statement in regard to your position as to whether agricultural cooperatives are justified.

Index